War's Waste

War's Waste

Rehabilitation in World War I America

BETH LINKER

The University of Chicago Press   Chicago and London

BETH LINKER is an assistant professor in the Department of History and Sociology of Science at the University of Pennsylvania.

The University of Chicago Press, Chicago 60637
The University of Chicago Press, Ltd., London
© 2011 by The University of Chicago
All rights reserved. Published 2011
Printed in the United States of America
20 19 18 17 16 15 14 13 12 11   1 2 3 4 5

ISBN-13: 978-0-226-48253-8 (cloth)
ISBN-10: 0-226-48253-7 (cloth)

Publication of this book has been aided by a grant from the Bevington Fund.

Library of Congress Cataloging-in-Publication Data

Linker, Beth.
  War's waste : rehabilitation in World War I America / Beth Linker.
      p. cm.
  Includes bibliographical references and index.
  ISBN-13: 978-0-226-48253-8 (cloth : alk. paper)
  ISBN-10: 0-226-48253-7 (cloth : alk. paper)   1. Disabled veterans—Rehabilitation—United States—History—20th century. 2. World War, 1914–1918—Veterans—Medical care—United States—History—20th century. 3. Medical rehabilitation—United States—History—20th century.   I. Title.
  UB363.L56 2011
  362.1086'970973—dc22   2010045280

♾ This paper meets the requirements of ANSI/NISO Z39.48-1992 (Permanence of Paper).

Contents

For Damon

The Roots of Rehabilitation

Watching Garth Stewart skip down the steps to catch a New York City subway, you would never suspect that he's wearing a prosthetic leg or that he's a veteran of the Iraq war. At the age of twenty, Stewart lost his lower left leg outside of Baghdad on April 5, 2003. After a brief return to service, he became an undergraduate at Columbia University with the help of the Servicemen's Readjustment Act, known informally as the GI bill, and received a bachelor's degree in history in 2009. His story has piqued the interest of many media outlets. He regularly receives invitations to gala events, hobnobbing with the cultural, military, and political elite. Former Deputy Secretary of Defense and friend Paul Wolfowitz calls Stewart "a rock star." Like any rock star, Stewart has big plans. He dreams of becoming a senator, perhaps even president one day. If he continues to follow the motto that he had taped to his dormitory door—"Rest When You're Dead"—he just might succeed.[1]

Stewart readily admits that he has become a "poster child of the war."[2] But he could not have achieved such heights if he had not first become a poster child of rehabilitation, triumphantly overcoming his disability to the point where he no longer appears disabled in the public eye. If a wounded soldier refuses rehabilitation and does not take the necessary steps to become as physically and financially independent as possible, he is considered a failure, not a hero. War wounds in themselves are not enough to earn respect. The maimed veteran who earns accolades is the one who makes good, applying his (and now her) military skills to fight for a full recovery. This is the ethic of rehabilitation.

That ethic has a history. From America's earliest wars to those of the second half of the nineteenth century, soldiers permanently injured in the line of duty received a lifetime of monetary reimbursement in the form of federal pensions, without any explicit demand of returning to gainful employment or a life of self-sufficiency.[3] Congress passed the first disability pension law in 1792, stating that all wounded soldiers of the Revolutionary War "shall be taken care of and provided for at the public expense."[4] At the end of the Civil War, President Andrew Johnson wrapped disability pensions in the blanket of patriotic duty, declaring that "a grateful people will not hesitate to sanction any measures having for their object the relief of soldiers mutilated . . . in the effort to preserve our national existence."[5] The pension system persisted for the remainder of the nineteenth century, operating on the assumption that in a just society, the government and its people owed disabled soldiers a simple debt of gratitude.[6] The disabled soldier had proven his patriotic worth. He had done his civic duty. Nothing more should be asked of him.

This would all change with the First World War.[7] Opposed to the perceived economic inefficiencies and political corruption of the pension system, Progressive Era reformers encouraged the Wilson administration to institute programs in rehabilitation, providing injured soldiers with long-term medical care and vocational training in order to drastically reduce—and potentially erase—cash payments made out to veterans. Emboldened by their faith in the new social and medical sciences, Progressive reformers believed that the government could (and should) "rebuild war cripples," curing them of their disabilities so that veterans of the First World War would make a speedy return to work and rely on their own wage-earning capacity rather than on government pensions. Men returning home with amputated arms and legs were to be "refitted," not only with prosthetic limbs, but also to an appropriate workplace, so that, as one rehabilitation advocate put it, a soldier's disability would not be a "handicap."[8] With proper physical and vocational reconstruction, blind men, deaf men, and dismembered men could all produce wealth and contribute to society.

Rehabilitation proponents aimed to rid the nation of "war's waste," a turn of phrase that referred to the human remains of war as well as to the economic cost that the nation had to endure after the battle was over.[9] Pension spending for the nation's aging Civil War veterans had reached an all-time high on the eve of America's declaration of war against Germany in April 1917.[10] One economist estimated that by the year 1916, the United States had spent over $5 billion on Civil War pensions, an amount that exceeded the price of the actual war.[11] Looking across the Atlantic at the human toll wrought by the First World War, Americans feared that engagement would lead to a heavier economic burden than all of the nation's previous wars combined. The promise of rehabilitation—that once the disabled soldier had been rebuilt, he would require no further monetary assistance—made the prospect of entering the Great War much easier to stomach.

But more than economic concerns drove the rehabilitation campaign. In World War I America, rehabilitation symbolized a dream, a hope that physical "handicaps," "pauperism," and "defects of manhood" could all be conquered on the home front.[12] It was a vision of physical and social perfectionism, but within limits. Unlike eugenicists who wished to sterilize "defectives" and enforce "ugly laws" to keep aesthetic undesirables out of public view, rehabilitation officials did not shun men with permanent disfigurements and missing limbs.[13] They sought instead to educate the nation to accept disabled soldiers while also providing the injured with the tools to reintegrate into "normal" life as seamlessly as possible.

Supporters of the World War I rehabilitation effort came from all walks of life. Some, such as Edward T. Devine and Julia Lathrop, had backgrounds in social work and others in law, for example, Judge Julian Mack. Certain big-business philanthropists like Julius Rosenwald supported the cause, but so, too, did labor leader Samuel Gompers. Presidents Theodore Roosevelt, William Howard Taft, and Woodrow Wilson all shored up the movement to get maimed soldiers back to work, as did key figures in the medical profession, such as William James Mayo, cofounder of the Mayo clinic, and Frank Billings, one-time president of the American Medical Association and dean of faculty at the Rush Medical College in Chicago.

Although a diverse lot, these reformers shared some common assumptions, most notably about the importance, virtue, and redemptive value of work.[14] Coming from native-born, middle-class white Protestant families, many rehabilitation proponents believed that work formed the core of a moral life. Following the old adage that "idle hands are the devil's tools," they thought that work could cure many evils, from alcoholism and adultery to nervousness and disability. Adherents of this philosophy included both the able-bodied and the disabled. Helen Keller, for example, insisted that work was necessary to living a happy life. Blind and deaf herself, Keller argued that a lack of wage work resulted in a "bondage of idleness and despair," "a state of idleness more terrible even than loss of sight."[15] The Protestant work-ethic informed the tradition of work therapy that began in the late nineteenth century as a treatment for both the insane and sufferers of tuberculosis—and it culminated in the early twentieth-century rehabilitation effort, pushing the work ethic to a new extreme, making work both the means and the end of recovery.[16]

While rehabilitation proponents considered the industrial workplace to be dangerous and potentially exploitative, they nonetheless believed that it was the best place for male laborers.[17] The workplace, as they envisioned it, promised a well-ordered life, in both the private and public spheres. If married men embraced their role as breadwinner, they would be busy at work, thereby securing a domestic life for their wives and children at home. Operating within the established structures of industrial capitalism—trying to improve what was already there rather than restructuring it—this breadwinner ideal motivated much of the labor legislation of the Progressive Era.[18]

Rehabilitation was thus a way to restore social order after the chaos of war by (re)making men into producers of capital. Since wage earning often defined manhood, rehabilitation was, in essence, a process of making a man manly. Or, as the World War I "Creed of the Disabled Man" put it, the point of rehabilitation was for each disabled veteran to become "a MAN among MEN in spite of his physical handicap."[19] Relying on the breadwinner ideal of manhood, those in favor of pension reform began to define disability not by a man's missing limbs or by any other physical incapacity (as the Civil War pension system had done), but rather by his will (or lack thereof) to work. Seen this way, economic dependency—often linked overtly and metaphorically to womanliness—came to be understood as the real handicap that thwarted the full physical recovery of the veteran and the fiscal strength of the nation.

Much of what Progressive reformers knew about rehabilitation they learned from Europe. This was a time, as historian Daniel T. Rodgers tells us, when "American politics was peculiarly open to foreign models and imported ideas."[20] Germany, France, and Great Britain first introduced rehabilitation as a way to cope, economically, morally, and militarily, with the fact that millions of men had been lost to the war.[21] Both the Allied and Central Powers instituted rehabilitation programs so that injured soldiers could be reused on the frontlines and in munitions in order to meet the military and industrial demands of a totalizing war. Eventually other belligerent nations—Australia, Canada, India, and the United States—adopted programs in rehabilitation, too, in order to help their own war injured recover. Although these countries engaged in a transnational exchange of knowledge, each nation brought its own particular prewar history and culture to bear on the meaning and construction of rehabilitation.[22] Going into the Great War, the United States was known to have the most generous veterans pension system worldwide. This fact alone makes the story of the rise of rehabilitation in the United States unique.

To make rehabilitation a reality, Woodrow Wilson appointed two internationally known and informed Progressive reformers, Judge Julian Mack and Julia Lathrop, to draw up the necessary legislation. Both Chicagoans, Mack and Lathrop moved in the same social and professional circles, networks dictated by the effort to bring about reform at the state and federal level.[23] In July 1917, Wilson tapped Mack to help "work out a new program for compensation and aid . . . to soldiers," one that would be "an improvement upon the traditional [Civil War] pension system."[24] With the help of Lathrop and Samuel Gompers, Mack drafted a complex piece of legislation that replaced the veteran pension system with government life insurance and a provision for the "rehabilitation and re-education of all disabled soldiers." The War Risk Insurance Act, as it became known, passed Congress on October 6, 1917, without a dissenting vote.[25]

Although rehabilitation had become law, the practicalities of how, where, and by whom it should be administered remained in question. Who should take control of the endeavor? Civilian or military leaders? Moreover, what kind of professionals should be in charge? Educators, social workers, or medical professionals? Neither Mack nor Lathrop considered the hospital to be the obvious choice. The Veterans Administration did not exist in 1917. Nor did its system of hospitals. Even in the civilian sector at the time, very few hospitals engaged in rehabilitative medicine as we have come to know it today.[26] Put simply, the infrastructure and personnel to rehabilitate an army of injured soldiers did not exist at the time that America entered the First World War. Before the Great War, caring for maimed soldiers was largely a private matter, a community matter, a family matter, handled mostly by sisters, mothers, wives, and private charity groups.[27]

The Army Medical Department stepped in quickly to fill the legislative requirements for rehabilitation. Within months of Wilson's declaration of war, Army Surgeon General William C. Gorgas created the Division of Special Hospitals and Physical Reconstruction, putting a group of Boston-area orthopedic surgeons in charge.[28] Gorgas turned to orthopedic surgeons for two reasons. First, a few of them had already begun experimenting with work and rehabilitation therapy in a handful of the nation's children's hospitals. Second, and more important, several orthopedists had already been involved in the rehabilitation effort abroad, assisting their colleagues in Great Britain long before the United States officially became involved in the war.

Dramatic changes took place in the Army Medical Department to accommodate the demand for rehabilitation. Because virtually every type of war wound had become defined as a disability, the Medical Department expanded to include a wide array of medical specialties. Psychiatrists, neurologists, and psychologists oversaw the rehabilitation of soldiers with neurasthenia and the newly designated diagnosis of shell shock.[29] Ophthalmologists took charge of controlling the spread of trachoma and of providing rehabilitative care to soldiers blinded by mortar shells and poison gas.[30] Tuberculosis specialists supervised the reconstruction of men who had acquired the *tubercle bacillus* during the war.[31] And orthopedists managed fractures, amputations, and all other musculoskeletal injuries.

Rehabilitation legislation also led to the formation of entirely new, female-dominated medical subspecialties, such as occupational and physical therapy. The driving assumption behind rehabilitation was that disabled men needed to be toughened up, lest they become dependent of the state, their communities, and their families. The newly minted physical therapists engaged in this hardening process with zeal, convincing their male commanding officers that women caregivers could be forceful enough to manage, rehabilitate, and make an army of ostensibly emasculated men manly again. To that end, wartime physical therapists directed their amputee patients in "stump pounding" drills, having men with newly amputated legs walk on, thump, and pound their residual limbs. When not acting as drill sergeants, the physical therapists engaged in the arduous task of stretching and massaging limbs and backs, but only if such manual treatment elicited a degree of pain. These women adhered strictly to the "no pain, no gain" philosophy of physical training. To administer a light touch, "feel good" massage would have endangered their professional reputation (they might have been mistaken for prostitutes) while also undermining the process of remasculinization. Male rehabilitation proponents constantly reminded female physical therapists that they needed to deny their innate mothering and nurturing tendencies, for disabled soldiers required a heavy hand, not coddling.[32]

The expansion of new medical personnel devoted to the long-term care of disabled soldiers created an unprecedented demand for hospital space. Soon after the rehabilitation legislation passed in Congress, the US Army Corps of Engineers erected hundreds of patient wards as well as entirely novel treatment areas such as massage rooms, hydrotherapy units, and electrotherapy quarters. Orthopedic appliance shops and "limb laboratories," where physicians and staff mechanics engineered and repaired prosthetic limbs, also became a regular part of the new rehabilitation hospitals. Less than a year into the war, Walter Reed Hospital, in Washington, DC, emerged as the leading US medical facility for rehabilitation and prosthetic limb innovation, a reputation the facility still enjoys today.

The most awe-inspiring spaces of the new military rehabilitation hospitals were the "curative workshops," wards that looked more like industrial workplaces than medical clinics. In these hospital workshops, disabled soldiers repaired automobiles, painted signs, operated telegraphs, and engaged in woodworking, all under the oversight of medical professionals who insisted that rehabilitation was at once industrial training and therapeutic agent. Although built in a time of war, a majority of these hospital facilities and personnel became a permanent part of veteran care in both army general hospitals and in the eventual Veterans Administration hospitals for the remainder of the twentieth century. Taking its cue from the military, the post–World War I civilian hospital began to construct and incorporate rehabilitation units into its system of care as well. Rehabilitation was born as a Progressive Era ideal, took shape as a military medical specialty, and eventually became a societal norm in the civilian sector.[33]

Despite the wide variety of war wounds and the highly individualized experiences of sustaining a permanent disability from battle, rehabilitation officials used amputee care as a model for creating a uniform treatment plan that could be implemented in all cases of disablement. Soldier-amputees, in turn, became (and still largely are) the model patients of the military rehabilitation effort. By war's end, amputee soldiers made up 5 percent of the wounded, whereas pulmonary tuberculosis accounted for 15 percent, and other orthopedic conditions 25 percent.[34] Although amputees represented only a fraction of the wounded, *War's Waste* focuses on this patient population in order to explain how and why they became the gold standard of rehabilitation. Antipension proponents who wanted to bolster the virtues of rehabilitation frequently utilized—in image and text—success stories of amputee veterans who, with the flip of a prosthetic strap, could appear cured. Artificial limbs allowed caregivers and society as a whole to engage in the illusion that the human ravages of war could be erased with a technological fix. The more complex and gruesome conditions—crushed skulls, empty eye sockets, permanent psychological trauma—brought the ideal of rehabilitation into question, for the aim of a cure appeared more uncertain and seemingly unattainable. Then, as now, rehabilitation holds out the promise that the wounds of war can be healed, and thus forgotten, on the national as well as the individual level.

The rise of rehabilitation was not simply a top-down process, controlled and orchestrated by the political and medical elite.[35] Disabled soldiers of the First World War engaged in power struggles with their treating physicians, playing a significant part as makers and shapers of the rehabilitation movement. These men fought against the rehabilitation ideal, which held that after six months of medical care, disabilities would disappear and wounded soldiers would be able once again to enjoy the rights and luxuries regularly taken for granted by able-bodied men. But while they fought against the propagandized ideal of rehabilitation, disabled soldier-patients agitated for more medical care, not less. In the absence of a generous pension system, many disabled soldier-patients wanted more time in the hospital to recover from their wounds, to adapt to prosthetic devices, and to overcome the physical and mental trauma of war. One of the outcomes of this resolve was that from World War I onward, US veterans of war would benefit from a federally funded system of health care, a form of socialized medicine that the nation's civilians would be denied time and again throughout the twentieth century. One of the greatest legacies of the Great War was the creation of the Veterans Administration and its system of federally funded hospitals, places devoted solely to the health care and rehabilitation of America's soldiers.[36]

By May 1919, six months after the signing of the Armistice that ended the Great War, approximately 120,000 disabled soldiers (almost the same number as war dead) passed through the army's rehabilitation program. Of those who underwent rehabilitation approximately one-quarter returned to duty, employed in some capacity by the US Army. Five percent received a surgeon's certificate of discharge for disability. Of that 5 percent, 90 percent of the disabled soldiers reportedly resumed their old occupations, without further medical care or vocational training.[37] Although less than 50 percent of disabled soldiers had completed their course of rehabilitation a year after the war's end, the program appeared to be a resounding success.

But the Progressive Era experiment in eliminating veteran pensions and bonuses came under repeated attacks in the decade following the end of World War I. Veterans of the Great War felt aggrieved by the federal government's attempt to scale back their benefits. By World War II, members of the American Legion and the Veterans of Foreign Wars reclaimed a more robust system of entitlements. The post–World War I agitation culminated in the GI Bill, granting all veterans paid college tuition to learn a new vocation, low-interest home mortgages, and a small "readjustment allowance" to cover unemployed veterans until they found work.[38]

Progressive Era reformers may have failed in their efforts to reduce the cost of veteran pensions and benefits, but the institution of rehabilitation fundamentally transformed how our nation would understand and treat its veterans. During the Great War, the United States was one of the few belligerent nations to make rehabilitation mandatory for all of its disabled soldiers.[39] If a disabled soldier refused rehabilitation, he faced the possibility of being dishonorably discharged and thus forced to relinquish all other benefits that come with military service. Through rehabilitation, Progressive Era reformers instituted a form of medical welfarism that obliges disabled soldiers to make every effort to recover from their war injuries so as to avoid becoming permanent wards of the state.

Garth Stewart is a product of this history. The leg amputation he sustained within the first months of the Iraq war has not slowed him down. He dons and doffs various models of prosthetic legs like clothing. He has one leg for boxing, another for running, and yet another for when he traipses around campus, like any other college student.[40] By his own account, he is always at work, whether in school, at fundraising events, or at the gym. He rarely sits still. The Progressive reformers who legislated and instituted rehabilitation could not have imagined a more ideal disabled soldier-patient. He is their dream made real.

The Problem of the Pensioner

On September 29, 1915, one hundred thousand visitors descended upon Washington, DC, for what was billed as the most exhilarating military parade in decades. Red, white, and blue adorned the parade route; American flags—now embroidered with forty-eight stars after the recent acquisition of Arizona and New Mexico—fluttered from windows and housetops. As throngs of people lined up from the Capitol Building to the White House, submarine torpedo boats navigated north up the Potomac River, docking in the Navy Yard near Georgetown Heights. The weather was "ideal," according to one *Los Angeles Times* reporter: a crisp autumn morning.

The participants in the parade were autumnal as well. This was not a dispatch of young American troops to the European theatre of war, the "crimson battle" that had been killing men by the hundreds of thousands since June 1914 (the United States would not commit to fighting the Great War until April 1917). Rather, this was a commemoration of Union war veterans. The gray-bearded "boys of '65" planned to retrace the path they took fifty years earlier, recreating the "Grand Review"—a parade that President Andrew Johnson arranged to celebrate victory over the Confederacy. Even by 1915 standards, the Grand Review was considered to be "one of the most spectacular military incidents of the history of the world." Approximately thirty thousand veterans marched in military formation along Pennsylvania Avenue.[1]

President Woodrow Wilson assumed the same position that Andrew Johnson did a half-century earlier on a grandstand in front of the White House. Born in Virginia and raised in Georgia and South Carolina, Wilson was the first southern-born president to preside over an anniversary parade of the Grand Review. At the sound of a 10:00 a.m. cannon shot, Wilson's Secretary of War Lindley M. Garrison, Secretary of the Navy Josephus Daniels, and the Commander-in-Chief of the Grand Army of the Republic Colonel David Palmer, joined Wilson on stage. Dressed in a conventional frock coat and silk hat, Wilson stood silently for four hours watching the procession of 30,000 elderly men (only a fifth of the original number of participants), who were, on average, 72 years of age. The men veered left and right toward cheering audiences, shaking hands, and gesturing for more applause. Eager to catch up with old friends, they talked "incessantly" along the parade route, at times drowning out the tried and true tunes of "When Johnny Comes Marching Home" and "The Battle Hymn of the Republic." When the disorderly platoons finally reached the grandstand, however, they "prettied up" to salute President Wilson and the other dignitaries. Wilson reportedly had tears in his eyes on and off throughout the procession.[2]

There were many reasons for Wilson to shed a tear on this day, for this dedication to the past seemed to portend the future. Only months earlier, the president, known for his commitment to peace, ordered Secretaries Garrison and Daniels to expand the army and navy, transforming the small 100,000-man US military into a world-class fighting power. Wilson's shift from neutrality to military preparedness evolved over the summer months of 1915, following the German submarine attack on May 7 on the *Lusitania,* a commercial British liner that sank, leading to the death of 128 Americans. Tensions on the diplomatic front, as well as within the Wilson administration, ran high. Secretary of State William Jennings Bryan, a well-known pacifist, resigned from office a month after the *Lusitania* sinking, convinced that war with Germany was in the offing.[3] The fact that Wilson wanted to expand the army and not just the navy gave Bryan all the proof he needed. Building a navy could be construed as a purely defensive move undertaken to protect American ships in enemy waters; expanding the army, however, indicated that the United States planned to send its own ground troops into trench warfare.[4]

Not all Americans who watched the semicentennial parade of the Grand Review felt the stirrings of patriotism produced by the military display of wars past or imminent. Despite the fact that in early twentieth-century America, the Civil War frequently "offered a mother lode of nostalgia," evoking sentimental memories of battle and glory in which romance often triumphed over reality, certain Americans found these veterans guilty of dragging the nation down a path toward moral and economic decline.[5] Contrary to the hero-worship displayed by the cheering crowds on the parade route, a growing number of early twentieth-century journalists, elected officials, social scientists, and politicians argued that the Civil War veteran who had once been honored for his sacrifice in battle was "now no better than [a] deserter, a straggler . . . a coward."[6]

The Civil War veteran's fall from grace arose not from antiwar sentiments, but rather from perceived injustices occurring in the Treasury Department, the branch of government responsible for paying out pensions to veterans after the war. By 1915, the aggregate cost of the 50-year-old pension system exceeded $3 billion dollars, with the US government paying out over $200,000,000 annually. Most Americans assumed that the cost of Civil War pensions would take a drastic plunge by the early twentieth century, since, fifty years after the war, very few veterans would still be living and able to make benefit claims. But the opposite happened. With the passing of each year of the twentieth century, more and more money went into the pension system. By 1915, the cost of the pension system had exceeded the actual cost of the Civil War itself.[7]

The men who benefited most from the pension system were precisely the men marching in the anniversary Grand Review parade—the men of the Union, the former soldiers of the Grand Army of the Republic (GAR). The pension system was originally established in 1862, a year after the War of the Rebellion commenced, as a way to recruit soldiers. Thereafter, the pension remained a prize granted exclusively to the victors of war.[8] Throughout the first half of the twentieth century, Confederate veterans were repeatedly denied Civil War pensions. Black veterans, whether Union or Confederate, faced a similar yet more dismal fate. The Pension Bureau engaged in explicit racism when it came to African-American claimants. Not only did fewer African Americans actually apply for pensions, but many who did were denied.[9] The majority of the pension money therefore went to white, GAR veterans who came from the mid-Atlantic and Midwestern region of the country. The home states of these beneficiaries—Ohio, Pennsylvania, Upstate New York, Illinois, Indiana, Michigan, and Missouri—collectively became known as "the pension belt."

In the years leading up to World War I, certain Progressive reformers began to see the veteran as a problem to the social order. This chapter tells the story of why Progressive Era reformers wanted to overhaul the pension system and how the call for reform set into motion the eventual institutionalization of rehabilitation. The primary criticism of the Civil War veteran welfare system was that it allowed pensioners to live off the government without the government expecting anything in return. The promise of rehabilitation was that it would get veterans, whether injured or not, back into the workforce, making them productive, tax-paying citizens. The reformers who criticized and effectively brought about the end of the Civil War pension system by 1917 were the same reformers who established institutions of physical and vocational rehabilitation as a substitute for pensions.

The World War I veteran returning home in November 1918 would, in many ways, walk a path different from that of the gray-bearded boys of '65 marching down Pennsylvania Avenue that September day. The notion of what the country owed its citizen-soldiers injured from war was radically redefined during the Wilson administration. The veterans of America's First World War were expected to become citizen-workers once their military service was over; they were to make useful lives, not to languish at the expense of the US Treasury. In a real sense, they were expected to be the opposite of the Civil War veteran. Thus, to understand why a program in rehabilitation was created in the first place, one must have an appreciation for how the Civil War pensioner became a problem that Progressive reformers aimed to solve.

Creating a Veteran Welfare State

At the turn of the twentieth century, the United States had one of the world's largest singularly targeted welfare plans, providing generous sums of money for its veterans of war. A century later, it is hard to believe that the United States ever outpaced Europe in welfare spending, but contemporary critics of the Civil War pension system were well aware of the fact. With the "Old World" serving as a benchmark for imperialism, socialism, and overtaxation, one early twentieth-century critic of the system noted, with disdain, that the United States spent twice as much on its veterans as did Germany, France, and Great Britain combined.[10] Such evidence led another critic to conclude—well before the Bolshevik Revolution—that America was "approaching not socialism, but communism in [its] pension measures."[11]

Yet the Civil War pension system was not intended to serve as a welfare-state program. Congress passed the General Law of 1862 because the Union needed more soldiers. The North paid a heavy toll during the first several months of war, and the pension plan was seen as a way to persuade men to volunteer for service instead of conscripting soldiers as the Confederacy was doing. Although this measure of political persuasion proved to be ineffective in raising a large enough army (the Union Conscription Act was instituted in March 1863), the General Law served as the baseline for the Civil War pension system well into the twentieth century.[12]

The fact that Congress instituted a pension program covering a war that was still being waged was unprecedented. The United States adopted its first veteran pension after the Revolutionary War.[13] By the 1810s, stories abounded of old war heroes being forced to "totter from door to door" with "desponding heart and palsied limb," begging for alms.[14] In the absence of a federal system of poor relief, as Great Britain had, President James Monroe in 1818 proposed a two-tier pension benefit based on military rank and financial need: officers received $240 annually while noncommissioned officers and privates collected $96 a year. To get the money, however, claimants had to provide testimony of service and an "oath of indigency" certified by a court of record.[15]

To certain congressmen, the Revolutionary War Pension Act of 1818 undermined the republic, creating a privileged class—here, white veterans—subsidized by taxes on the poor. Others argued that it flew in the face of "the republican ideal of the citizen soldier who serves his country out of a sense of duty rather than for material gain."[16] Republican senator Nathaniel Macon from North Carolina warned that, in addition, it would set a dangerous—and costly—precedent. "To undertake to provide for those who will not provide for themselves," Macon argued, "will, on experiment be found an endless task . . . it will drain the treasury, no matter how full."[17] Supporters of the pension countered with assurances that it would be a short-lived program that would apply to fewer than 2,000 men. When the bill was eventually passed on March 18, 1818, the Senate estimated that it would cost approximately $115,000.

Such forecasts proved naïve. Within the first six months of the bill's passage, the War Department received over 20,000 applications.[18] A year later, the cost of the program amounted to nearly $2 million, seventeen times more than predicted. Despite repeated objections, the pension law persisted, altered only by courts applying stricter forms of means-testing in order to weed out the thousands of wealthy applicants who submitted fraudulent claims. With more stringent means-testing in place during the 1820s, the Revolutionary War pension system became, according to historian John Resch, a "quasi-poor law," making the federal government the almoner for the country's indigent veterans and their dependents.[19]

In an attempt to avoid the pitfalls of the past, the framers of the 1862

General Law adopted a different system of means-testing, one based on degree of disability, rather than on poverty. Thus, those who created the General Law instituted a system heavily reliant on expert testimony to establish a claimant's honesty. Whereas supporters of the Revolutionary War pension believed that veterans should not have to be subjected to the "humiliating [process] of . . . producing surgeons' certificates" attesting to a claimant's physical state of being, the General Law insisted that medical evaluations were necessary to keep the system free from fraudulent claims.[20] The end result was a complex filing process that relied on physician affidavits, personal testimonies, notaries, and lawyers.[21] The Pension Bureau—a federal department that grew to be one of the largest employers in Washington, second only to the US Postal Service—would then review and process the claim.[22]

At the outset, the General Law was more comprehensive and broader in scope than the 1818 system, providing compensation to war widows, dependents, and veterans disabled by injury and disease. Indeed, for the first time in the nation's history, men disabled from chronic disease, such as tuberculosis, were eligible for pensions. Benefits were further expanded in March 1865, just two weeks before Robert E. Lee surrendered to the Union, when Congress established the National Asylum for Disabled Volunteer Soldiers (later renamed the National Home for Disabled Volunteer Soldiers). Before national homes had been established, local philanthropic organizations took responsibility for helping injured Civil War soldiers recuperate. In the North, middle- and upper-class women established privately run soldier's homes in almost every major urban center.[23] Yet by war's end, the numbers of the returning injured put too much demand on the private sector, forcing the creation of a nationalized institution for disabled soldiers.

Put together, the General Law and the National Home for Disabled Soldiers moved a country that was historically suspicious of consolidated federal power into a full-fledged—albeit highly selective—welfare provider. According to historian Patrick J. Kelly, during the first two decades of its existence, the National Home provided shelter and care to one out of every twenty men who served in the Union forces. And it did so, according to Kelly, with little public resistance. "At this institution," he writes, "Americans witnessed with great equanimity and barely a word of protest, a significant example of the state's capacity to assume a social welfare responsibility previously accepted on the local level."[24]

Support for soldiers' homes and the General Law was primarily found in the North—the region that benefited most from such measures—and within the Republican Party. Throughout the Gilded Age, the Republican

Party was notorious for the campaign tactic of "waving the bloody shirt" in order to activate wartime memories of the North's victory over the South, heightening sympathy for Union soldiers who struggled to save the nation. In short, the bloody shirt was a way for Republicans to demonize Democrats as the party of traitors.[25] At the same time, the GAR evolved into a powerful political machine, an organization that lobbied Congress on behalf of Union veterans. The GAR could make or break an election, as long as Union veterans voted en bloc.

The revenue for the Civil War pension program came from tariffs on imported goods. The Republican Party favored high tariffs as a way to stem marketplace competition coming from foreign countries, keeping domestic industrial employment steady and wages high. While such tariffs worked to the benefit of northern manufacturers, they hurt southern cotton producers and western agriculturists. Because of Republican commitment to tariff protectionism, Democrats accused the GOP of being nothing more than the tool of northern capitalists. Indeed, certain Democrats believed that the Republican Party, contrary to its espousal of honoring Union veterans who sacrificed their lives for the sake of the country, used the Civil War pension system to further the tariff agenda, siphoning off unseemly surpluses that accumulated in the US Treasury from high tariffs.[26]

Considering how regionally defined and circumscribed the General Law was, one would think that most vocal resistance would come from southern politicians, but this did not prove to be the case. Despite their reputation for fiscal conservatism, the Redeemers in Georgia began providing pensions for disabled Confederate soldiers in 1875, authorizing each county of the state to pay maimed and indigent soldiers $100 annually. Virginia established its own soldier's home, the Lee Camp Home, appropriating approximately $285,000 in state funds during the years 1884–98. In 1879, North Carolina became the first state to make a permanent tax-based provision for veteran pensions.[27]

Although much of the post-Reconstruction South was "hobbled by debt and limited by a small tax base that was constrained by the region's industrial underdevelopment and stagnating prices for the region's chief agricultural commodities," the state-by-state veterans pension systems that cropped up during the 1870s and 1880s enjoyed political popularity, for they demonstrated that the South could take care of its own and, at the same time, bolster the Lost Cause.[28] Indeed, later in the century, when certain congressmen argued that Confederate soldiers should be subsumed under the General Law, receiving the same benefits as Union veterans, a Tennessee veteran's organization argued that such an act would injure the pride and self-esteem of the Confederate soldier. Or as one Georgia Confederate veteran put it: "To make our gallant Confederate veterans dependent for their future . . . upon piteous charity of a former foe, would not be in keeping with the southern manhood so . . . heroically illustrated."[29]

Mugwump Opposition

The first wave of protracted public protest against the pension system in post bellum America came from northerners within the Republican Party itself. An independent wing of the Republican Party—dubbed the "Mugwumps" by party loyalists—came out in strong opposition to tariff protectionism and veterans' pensions.[30] The Mugwumps earned their name during the 1884 presidential election, when they gave their support to Democrat Grover Cleveland over Republican nominee James G. Blaine. For over a decade leading up to the 1884 election, the Republican administrations under Grant, Hayes, Garfield, and Arthur continuously reinterpreted the General Law to the GAR's advantage. The 1873 Consolidation Act, for example, made it possible for veterans to be compensated "for conditions and diseases contracted during military service that *subsequently* resulted in disability." This piece of legislation made it possible for veterans who developed conditions such as rheumatism or tuberculosis *after* the war to receive compensation. The Consolidation Act proved significant, for by 1888, 64 percent of all pensions covered diseases and conditions not sustained on the battlefield.[31] After the Consolidation Act, the General Law was then further liberalized under President Hayes in 1879, when he signed into law the Arrears Act. With this law, veterans who had never before applied for a pension could do so with the assurance of receiving a lump-sum back payment covering the preceding fourteen years since the war's end. The average first payment to arrears pensioners was $953.62, a considerable amount of money given the fact that the average annual money earnings of nonfarm employees at the time amounted to approximately $400.[32]

Mugwumps supported Cleveland for several reasons, one of which was his ardent opposition to the Republican record on pensions. Mugwumps fashioned themselves as gadflies to the Republican Party. As college-educated, elite businessmen, clergy, educators, and literary editors from New York and New England, they had the means to make their views known. Writing in 1885 for *Harper's Weekly*, editor-in-chief George William Curtis defended his Mugwump position as one that stood on a moral high ground, above partisan politics. Mugwumps, he wrote, "are intent to show young men who are disposed to obey their conscience rather than the party caucus what they have to expect if they do not obediently run with the machine." "The young Republican or Democrat," Curtis continued, "may wisely remember . . . that it is better to regard a party as a servant, rather than himself as the slave of the party."[33] The 1884 Mugwump vote, then, was not an affirmation of the Democratic Party, but rather a symbol of protest against a Republican Party that was seen to be engaged in "spoils system" politics, unfair taxation, and cronyism.

During his first term in office, Cleveland vetoed 228 pension bills, two-thirds as many as had been signed by all of his predecessors. Most boldly, he vetoed the Dependent Pension Bill of 1887, which proposed to cover all disabled veterans, even if the injury did not occur as a result of military service.[34] The Dependent Pension Bill came out of GAR agitation for a universal service-pension, legislation that would grant pensions to all Union veterans, regardless of whether or not they were disabled. The Dependent Pension Bill was a form of appeasement, a way for Republicans to maintain GAR support without going to the extreme of granting universal pensions with no system of means-testing in place.

Among the GAR, Cleveland's hard line against pensions earned him the reputation as an "inhuman monster."[35] While the veto received warm praise from Mugwumps, it had the opposite affect on GAR propagandists, who used the event to demonize the Democratic Party and galvanize the veterans' vote for Republican candidate Benjamin Harrison in the 1888 presidential election. Harrison's win—widely attributed to the "soldier's vote"—rested on the campaign to "pension the boys who wore the blue."[36]

Harrison fulfilled his promises. A little over a year after taking office, he signed into law the Dependent Pension Bill—the one that Cleveland had vetoed. The 1890 Dependent Pension Act changed the pension system in several crucial ways. First, it severed the link between service-related injuries and pay. Moreover, it reduced the length of military service needed to qualify for pensions to ninety days. Lastly, rear-line manual laborers who never witnessed, let alone engaged, in battlefront combat were now eligible to receive federal money as long as they could prove that they sustained a work-related injury while in military uniform.[37] Anticipating Harrison's passage of the Dependent Pension Act in 1890, Harvard University president Charles W. Eliot warned his fellow Mugwumps that without the requirement of a service-related wound, Americans would no longer be able to "tell whether a pensioner of the United States [was] a disabled soldier or sailor or a perjured pauper who [had] foisted himself upon the public treasury."[38]

With a pension-favoring Republican back in office, a second round of Mugwump complaints filled the nation's opinion magazines. Leaders of the Social Gospel movement, Washington Gladden and Francis G. Peabody, joined forces with publicist Charles Dudley Warner (famed co-author of *The Gilded Age: A Tale of To-Day* with Mark Twain) and economist Richard T. Ely to form the "sociological group," white male professionals who believed the pension system was an affront to social justice.

The sociological group did not oppose government intervention, per se—they did not align themselves with the Bourbon Democrats' platform of states' rights. As leaders of the Social Gospel, Gladden and Peabody both pushed for an expanded government, believing that it, like the church, could bring about the Kingdom of God here on earth. In 1885, Ely, then professor of political economy at Johns Hopkins University, helped found the American Economic Association (along with Gladden and future president Woodrow Wilson) to repudiate laissez-faire economists and their noninterventionist stance in matters of property and contract rights. Ely remained a social reformer throughout his career, becoming the first president of the American Association for Labor Legislation (AALL) in 1906.

The problem of the pension system, the sociological group maintained, was that it created a privileged class, the GAR, on the backs of hard-working nonveteran working-class laborers. "What masquerades to-day" as social justice, wrote the sociological group in 1891, "is simply the distribution to one class in the community of what belongs to another," an example of "inequality through taxation."[39] In short, what helped the veteran hurt the working-class man. Before the era of workmen's compensation, the Union veterans' pensions program was the only federal entitlement that offered income assistance.

One way to solve the perceived problem of inequality would have been to expand the pension system to provide universal disability coverage for both veterans and nonveterans. But members of the sociological group opposed this kind of expansion, for such an arrangement smacked of socialism and unwieldy paternalism. For all their reformist intentions, members of the sociological group were clear to distinguish themselves from the "militant" socialists of Europe. The United States, having fought for its independence, was supposed to have learned from "the disastrous example of Europe . . . with [its] discordant interests, . . . perpetual wars and enormous armaments, taxing every man, woman and child." Rather than being a studious pupil, the United States appeared to be a reckless troublemaker when it came to veterans' pensions, as politically corrupt as the Old World. "Prussia under Frederick the Great" the sociological group wrote, "distributed annually to disabled veterans less than one week's expenditure in the United States at present." The numbers did not stop there. The sociological group figured that "62,000,000 people in the United States are annually paying $44,000,000 for a military establishment, $22,000,000 for a navy, and $160,000,000 for pensions—a total of $226,000,000 which is 80 per cent of what the combined 86,000,000 people of France and Germany together pay for their armaments."[40]

With numbers like these, critics of the pension system claimed that the United States was no longer on the verge of socialism but in it. The Mugwumps blamed Republicans, such as Harrison, for emptying the coffers of the federal treasury in order to secure the soldier's vote. Self-interested politicians and the GAR were bringing about a decline in "American virtue," pension critics maintained, allowing petty temptations, state centralization, sentimentality of the war hero, and dishonesty to flourish. If left unchecked, warned the sociological group, the pension system would "climax . . . in communism."[41]

Tapping into the tradition of American exceptionalism helped the Mugwumps to paint a grim picture of the pension system. In their hands, the Dependent Pension Act looked liked a radical departure from the country's history—in a word, it looked un-American. This line of argument applied not simply to the system, but to pensioners themselves. Because the Dependent Pension Act enlarged the pool of potential recipients, making it possible for injured rear-line soldiers to receive benefits, Mugwumps like Eliot believed that such widespread dependency would have a corrosive effect on the rest of the US citizenry, chipping away at the American ideal of individual autonomy and economic independence.

At stake for critics of the Dependent Pension Act was the fate of American "character," in particular, American manhood. The sociological group went so far as to argue that all pensions, of any kind, were emasculating. This was particularly troublesome since the very group who received pensions were supposed to be the pinnacle of masculinity, soldiers of war who were revered for their courage, honor, and duty to the country. Injured veterans begging for alms on city streets posed a problem not only for urban cleanup, but also for the education of the young. "The old soldiers, independent [and] self-respecting," wrote the sociological group, "should be a strong force in the community, an example and inspiration to us, to our children, and children's children."[42] The concern was clear: If the most manly of men became dependent on the state, then what was to keep the rest of the country—the "unmanly" men and women—from expecting the same?

Despite such criticism, the Dependent Pension Act remained law throughout the 1890s, surviving another Cleveland presidency (1893–97) and persisting into the twentieth century. The percentage of Civil War veterans receiving benefits ballooned over the decade, going from 39 percent in 1891 to 74 percent in 1900.[43] Half of the government's federal budget was devoted to pensions by 1900. In the face of mounting costs and a persistent GAR lobby asking for a universal service-pension, Theodore Roosevelt issued Executive Order 78 in 1904, granting pensions to all veterans over the age of 62, declaring that old age itself was a disability.[44] Although at first taken to be a cost-cutting measure, ridding the pension system of administrative expenditures devoted to means-testing, Roosevelt's expansion led to even greater costs, for a vast majority of Civil War veterans were (or were very nearly) aged 65 and over by 1905. In certain northern states, the proportion of pensioners climbed to one-fifth of the population of elderly citizens by 1910.[45]

Antipension Progressives

What began as a small group of Mugwump dissenters turned into a full-fledged Progressive Era reform movement against the Civil War pension system by the 1910s. Leading the charge was journalist and publicist Walter Hines Page. In many ways, Page proved to be an ideal spokesman for reform of the pension system. Born in 1855 in Cary, North Carolina, Page came from a family ambivalent about the Confederacy. His father, Allison Francis Page, never fought on behalf of the Stars and Bars. After establishing his own newspaper, the Raleigh *State Chronicle,* in 1883, the young Walter permanently relocated to the North at the age of thirty, first moving to New York and then to Boston to become the first southern editor-in-chief of the *Atlantic Monthly* in 1898. While in New York, Page moved in Mugwump circles, working for the *New York Evening Post*—"the altar of the Mugwump creed"—and participating in the Reform Club of New York, a private society that helped put Grover Cleveland into office by publishing inexpensive, ready-to-print pamphlets touting the benefits of lower tariffs.[46]

In 1900, Page moved back to New York from Boston to team up with Frank N. Doubleday to found the publishing house, Doubleday, Page and Company. Here, Page started his own magazine, *World's Work,* a nonfiction monthly that emphasized public affairs. With *World's Work,* Page created a new kind of magazine, different from the *Atlantic* and *Harper's* in that there was no literary section, and distinct from the big-circulation periodicals of the 1890s (*Saturday Evening Post* and *McClure's*) in that there was little entertainment value. In order to boost sales, Page came up with the idea of an investigatory series—stories that would appear in installments over several consecutive issues of the magazine, keeping his readers buying from one month to the next.[47] At twenty-five-cents an issue, *World's Work* grew from 16,000 subscriptions in 1901 to a circulation of 100,000 by 1907, a number that made the magazine competitive with other leading magazines at the time. The magazine's claim to fame was its 1906 series uncovering the unsanitary conditions found in the nation's meatpacking industry, timed in conjunction with Doubleday's release of Upton Sinclair's book, *The Jungle.*

In addition to its exposés on urban schools, railroad safety, and municipal corruption, *World's Work* served as the mouthpiece for the growing public dissatisfaction with the pension system. In some of the magazine's earliest editorials, Page attacked the pension system head on, claiming that the nation's reputation had been sullied by a "body . . . of mendicants, who profit[ed] by the nation's generosity and by the weakness of [the American] political system."[48] Page opposed the pension system for many reasons, but none so much as the fact that it promoted sectionalism. From an early age, Page was committed to reconciliation between the North and South. He refused to believe that there was a separate "southern problem." To him, the South's "troubles were variations of general American conditions, worsened and complicated by economic and cultural backwardness, racial tensions, and misguided notions left over from the Lost Cause."[49] Contrary to proponents of the Lost Cause, Page thought that southerners should forget the past and look to a future, toward a "new" South built not upon industrial capitalism and northern investment, but rather upon educational and agricultural reform.[50]

The pension system, Page believed, kept the memory of the Civil War alive in a way that was deleterious to the South and to the nation as a whole. "The Civil War," he wrote in 1904 just as Roosevelt put forth Executive Order 78, "has passed out of the minds of most men less than fifty years of age—or it would pass out if the promoters of pensions would decently permit them to forget it."[51] With increasing liberalization of the federal pension system, however, the disparity between the haves and have-nots became ever more apparent. In 1907, economist William H. Glasson, who taught at Trinity College in North Carolina, began to accumulate figures to highlight such disparities. Among the northern states,

Ohio and Indiana received twice as much pension money proportional to their numbers of veterans as did Kansas and Vermont. The discrepancy between North and South, however, was even more striking. Whereas Maine, New Hampshire, and Vermont together received $5,800,000 in pensions in the year 1910, South Carolina—with approximately the same population as the three New England states combined—got $292,000. Assuming that the "single Southern state contributed equally with the New England trio to the federal taxes," Glasson wrote, "the New Englanders enjoyed an advantage for that year of $5,500,000 in the expenditure of the federal revenues for military pensions."[52]

While Page and his staff of reporters lambasted the pension system, GAR supporters continued their fight to establish a universal service-pension for all Civil War veterans, regardless of age, disability, or length of service. During the first and second decade of the twentieth century, Congressman Isaac Sherwood from Ohio became the most visible proponent of instituting a pension that would require no means-testing, not even age restrictions. Sherwood was a man of letters himself. After finishing law school in Ohio in 1857, he purchased a weekly newspaper, the *Williams County Gazette,* in Bryan, Ohio. An ardent abolitionist, Sherwood responded to Lincoln's call to arms, enlisting as a private for the Union Army in 1861, ascending in rank to Brigadier General by the war's end. After the war, he continued his career in journalism and also became involved in Ohio politics, serving as a Republican representative to Congress in 1873–75. After his term, he abandoned the GOP, opposing its commitment to the gold standard and in 1879 ran as a gubernatorial candidate for the Ohio Democratic Party.[53]

At the age of seventy-one, Sherwood returned to Congress in 1906, this time as a Democrat and with a commitment to getting a universal service pension bill passed through Congress once and for all. As chairman of the Committee on Invalid Pensions in the Democratic House of Representatives in 1911, Sherwood proposed a bill that would pay pensions on a graded scale according to length of service, disregarding age or injury in determining the rate of pay. While those who served ninety days would receive $15 a month, the veterans who fought in the war for one year or more would get a monthly pay of $30, allotting them a dollar a day. In 1912, Republican President William Howard Taft signed a variation of Sherwood's pension bill into law.[54]

"Dollar-a-Day" Sherwood, as he became known, created outrage among Page and the staffers at *World's Work*. "There is a story current in Washington," wrote Page, tongue in cheek, "that when the Sherwood Pension Act passed Congress the burglar alarm in the Treasury Department went off."[55] The act proved to be expensive indeed. Since over 70 percent of Civil War survivors had served one year or more, the pension roll became even longer. According to one figure, the yearly amount paid for pensions increased from \$152,986,000 in 1912 to \$172,409,000 in 1914.[56]

That Sherwood came from the Democratic Party caused much soul searching among those who still glorified President Cleveland as the only real politician who fought against the corrupt pension system. Until Sherwood, every major act that liberalized the pension system came from the GOP, the party that—aside from Cleveland—had controlled the presidency since war's end. But Sherwood proved to be the exception to the rule. "In Isaac R. Sherwood," Page lamented, "the Democrats have produced a pension fanatic who has no counterpart in the long list of Republican war horses."[57]

For men like Page, the Sherwood Act solidified a political commitment to "the Progressive Programme," reflecting the attitude that both the Democratic and Republican parties were victims of corrupt machine politics. Progressive reform was well underway by 1912, with policies enacted across the country regulating corporations and improving working conditions. But, according to Page, the Progressives had not yet worked up the nerve to tackle the pension system. In a 1912 editorial that was meant to prod Progressives into taking a stance on the matter, Page wrote:

It is idle to talk about the free government of the people and about efficient and clean use of public money till this scandal is attacked and its mendicant organization is broken up. A little Congressional or Presidential *courage* would cause it to fall to pieces. Yet this Congress passed [the Sherwood] bill, a Democratic Speaker ostentatiously voting for it and the Republican President unquestioningly approving it, which made an indefensible addition of twenty-five millions or more a year to the present pension budget.[58]

It is telling that Page employed the rhetoric of courage in his indictment against the Progressives, for he belonged to a generation of men in the throes of a masculinity crisis. This was an era when men from all walks of life seemed to be suffering from insecurities concerning their manhood. Unemployed working-class men, still reeling from the economic depression of 1893, worried about their ability to provide for their families. Wealthy industrialists and financiers felt less autonomous than ever with growing governmental restrictions. Middle-class white-collar professionals believed that their desk jobs were making them physically soft. Immigration, the closing of the frontier, the "enfeeblement of liberal Protestantism," the enfranchisement of women, and the changing industrial workplace all seemed to be in collusion, bringing about the demise of American manhood.[59]

One solution to the masculinity crisis was to glorify war. This was the tactic of men such as Oliver Wendell Holmes, Theodore Roosevelt, and G. Stanley Hall, who clamored for a military conflict, thinking that battle would rekindle the days of fraternity among men, when heroism mattered more than commercial acquisition, when soldiers understood the value of suffering and the "strenuous life." It was this kind of thinking, historian Kristin Hoganson tells us, that prompted the US-led invasion of the Philippines, resulting in the 1898 Spanish-American War. For the men who turned to state-sanctioned violence as a way to bolster the nation's sense of virility, the aging Civil War veteran became part of the cause. "Because American men commonly associated the civic virtue necessary for democracy with the manly character exemplified by soldiers," Hoganson argues, "the dwindling tally of Civil War veterans led a wide range of men to fear that unless the nation forged a new generation of soldier-heroes through war, US politics would be marked by divisiveness, corruption, and weakness."[60]

Page's call for political courage in tearing down the pension system, then, was an attempt to redefine the terms of political masculinity, wresting it from the hands of prowar types like Roosevelt who seemed, by virtue of their battle experience, to define manhood. This was not an easy task, for Page, being part of the generation too young to have fought in the Civil War, was up against not only decorated Union veterans such as Sherwood but also the bellicose types of his own generation who rushed headlong into the Spanish-American War. Ever since their Mugwumps days, critics of the Civil War pension system had come under attack for their lack of manhood. In 1896, Republican representative Henry R. Gibson argued that those in his party who broke ranks to vote for Cleveland were effeminate. "A man who has not enough zeal to make a partisan," he concluded, "is not apt to . . . make a patriot."[61] Other Republican Party loyalists derided the Mugwumps by calling them "eunuchs," and "political hermaphrodites." To Republicans like Gibson, Mugwumps failed the test of manhood on two counts. First, by undercutting the fraternalistic, military-style tradition of politics, the Mugwumps favored a more female-style of political engagement that preferred education over brotherly loyalty. Second, to attack the pension system was to degrade the nation's greatest symbols of manhood. In a speech delivered to Congress in 1900, Gibson urged greater liberality when it came to Civil War pensions, saying that "the old soldiers [are] the saviors of the nation . . . and their pensions shall be paid with a generous hand and a smiling face [so] that the pensioner should feel that his country appreciated him for his sacrifices and honored him for his bravery and patriotism."[62]

Part of *World's Work*'s tactic in bringing the pension system into question was to turn the military ideal of manhood on its head, portraying the war veteran not as brave hero, but as a helpless, feminine dependent. Take, for example, a *World's Work* 1910 article "The Pension Carnival," the first multi-issue exposé on the pension system written by the famed *New York Times* reporter William Bayard Hale. "Remember the halos of romance that encircled the one-armed or limping heroes," Hale begins, "who told never-wearying stories of Antietem . . . [and] the wilderness of Gettysburg?" "The Pensioner" Hale continues "was indeed a hero . . . a man who had seen carnage and had wrought it." But today, he concludes, "the pensioner is a suspect. The common presumption is against his being a hero. The presumption, cynical perhaps, but not unjustified, is that he is as likely to be a crook or a hustler or a peddler."

The Spanish-American War brought the pension debate to a head, with pension critics feeling as if they had the upper hand against their propension opponents. Known as the "splendid little war," the Spanish-American War was short in duration, resulting in relatively few causalities, especially in comparison to the Civil War. With only roughly 300 US soldiers lost in combat (approximately 3,000 died of disease), the tactic of waving the bloody shirt on behalf of the Spanish-American-War veteran appeared preposterous to pension critics.

Pension critics found the Spanish-American War veterans' sense of entitlement most irksome. In his 1915 *World's Work* article, "Pork-Barrel Pensions," Burton J. Hendrick put it the following way: "Pensions had become part and parcel of our national psychology. The young men who enlisted for the Spanish war were in many cases the sons of Civil War veterans; in any event they were familiar with our national pension habit; in their minds, serving one's country, even serving it bloodlessly, necessarily implied a pension. Only such a mental attitude can explain the eagerness with which these survivors began to attach themselves to the pension list."[63]

This was a blistering indictment against those who claimed that the Spanish-American War was good for the country, producing stronger, more courageous men. Nothing seemed more effeminate, Hendrick proposed, than a veteran who never engaged in combat on an actual battlefield asking for a lifetime of pension pay from his government.

In addition to their gender-laden attacks, pension critics of the post–Spanish-American War era argued that the continuation of the Civil War

pension system for all subsequent wars was fiscally irresponsibl
twenty years after the Spanish-American War had ended, appre
29,000 of the 30,000 soldiers who saw active service in Cuba v
ing pensions.[64] By 1916, this group of veterans had drawn
$53 million from the Treasury. Belonging to a cohort of me:
an undo burden of continuing to pay for a war (the Civil War
never witnessed or fought themselves, early twentieth-centu., r
critics did not want the same to happen to future generations. The great
scandal of the day was not the current cost of the Spanish-American War,
but rather the aggregate cost of the Civil War. Now fifty years past, the
War of the Rebellion cost the United States over $5 billion dollars in total
pension distributions, exceeding the price of the actual war.[65]

Pension critics such as Hendrick believed that, if anything, the Spanish-
American War should serve as a cautionary tale for future policies concerning veterans' benefits. "In 1948, fifty years after the fall of Santiago,"
Hendrick portended, Spanish-American "veterans will be drawing about
$33 million a year from the Federal Treasury; our total expenditures, in
the course of a half a century, will reach nearly $1 billion."[66] Hendrick
was not far off the mark. In a 1950 Brookings Institute report on veterans'
benefits, author Mildred Maroney reported that pensions for Spanish-
American War veterans had exceeded $1 billion.[67]

With Europe fighting one of the deadliest and costliest wars in history,
and the US's engagement becoming ever more likely, these figures gave
certain lawmakers pause. "If pensions for the Civil War led to extrava-
gance and fraud that will probably run riot until about 1960," wrote one
critic, "the very thought of such a pension system, for this, [the world's]
greatest war is staggering."[68] Looking across the Atlantic at the human
toll wrought by the Great War, Americans did not need expert predic-
tions to know that involvement may bring about a heavier economic
burden than all of the nation's previous wars combined.

From Pensions to Rehabilitation

The First World War brought about a radical change in the way US veterans' benefits would be disbursed. Soon after America's declaration of
war, President Woodrow Wilson along with his Secretary of the Treasury William McAdoo, created the Section on Compensation for Soldiers
and Sailors, within the Council of National Defense's Advisory Committee on Labor, chaired by Samuel Gompers. That Wilson and McAdoo
turned to the Committee on Labor, rather than to the Pension Bureau,

cated their desire to build a veterans' compensation system anew, one that would treat soldiers more like laborers than valorized men of the military.[69]

Indeed, the goal of the Wilson administration was to reformulate veterans' benefits by using workmen's compensation laws—adopted by some 40 states at the time—as a model. To achieve that end, McAdoo appointed Judge Julian W. Mack, a Chicago federal circuit court judge, to chair the Section on Compensation for Soldiers and Sailors. Mack was a leading figure in the Progressive reform movement. A Chicago resident, he developed strong relationships with Jane Addams and Children's Bureau director Julia Lathrop, who would later assist Mack in drafting the new system of veterans' benefits.[70] While serving as magistrate of Chicago's Cook County juvenile court, Mack became known as a child-welfare reformer, insisting that every attempt be made to rehabilitate juvenile offenders rather than incarcerating them. As vice-chairman of the White House Conference on the Care of Dependent Children in 1909, he maintained that the lack of workplace safety and low wages were the leading causes of destitution among urban immigrant families and children.[71]

To come up with a new plan for soldiers' benefits, Mack solicited proposals from the nation's leading social insurance advocates of the day. Henry R. Seager and Samuel McCune Lindsay, founder and member of the AALL, respectively, both submitted proposals arguing for "war insurance." Just as unemployment, disability, and health insurance indemnified wage workers against the risks of modern industrial production, so too would war insurance protect soldiers and their families against the risks of battle and the loss of their civilian income.[72] "During this war," wrote Lindsay, "the government of the United States is the employer and . . . the people . . . are the consumers . . . for whom the operations of war are carried out." "The government therefore," he continued, "should bear the whole cost of compensation for death and disability for officers . . . [and] enlisted men, and distribute the burdens through taxation."[73] Wanting to apply workmen's compensation at a federal level, Lindsay believed that soldiers should have a contractual agreement with the government, just as industrial workers had with their employers.

Making their preference for insurance clear, the Section on Compensation for Soldiers and Sailors recommended that responsibility for veterans' benefits be moved out of the Pension Bureau—the federal agency instituted under the 1862 General Law—and given to the newly established War Risk Insurance Bureau. Created in September 1914, the War Risk Insurance Bureau provided insurance against war-related loss or damage to US merchant ships that conducted business in European waters.[74] The plan of the Section on Compensation for Soldiers and Sailors was to extend this kind of insurance to the "human material" of war, insuring the bodies of soldiers sent into war. The move, in essence, took veteran's legislation out of the hands of interest groups and congressional lawmakers and put it into the hands of Progressive reformers and social science experts interested in instituting a system of social insurance at the federal level.

Despite some criticism from the traditional guardians of the pension laws, the War Risk Insurance bill put forth by Mack and his committee passed Congress on October 6, 1917, without a dissenting vote, making the US government one of the largest insurers in the nation. The bill passed Congress with relative ease because, as a new piece of veterans' legislation that applied *only* to World War I veterans, it did not threaten the Civil War pension system. The Wilson administration, in other words, never brought about an end to the Civil War pension system, but rather reset the clock by establishing new rules that would apply to the veterans of the First World War.[75] Because the legislation did not bring an end to the Pension Bureau or the General Law, northern Republicans were able to vote in favor of the measure without worry about losing GAR votes. At the same time, southern Democrats supported the bill because it offered, for the first time since 1865, federal support to veterans from the former Confederacy.

Pension critics welcomed the War Risk Insurance Act (WRIA) with glee. *Outlook* magazine heralded it as "the greatest safeguard of mankind."[76] *The Nation* reassured its readers that with War Risk Insurance in place, "the old degrading and extravagant system of pensions is completely abolished."[77] *World's Work* described the act as "honest, efficient, patriotic, and businesslike," preventing future veterans from becoming "parasites" on the federal government.[78] Framers of the bill also touted its importance to the history of veteran welfare. In a 1918 address delivered to the American Philosophical Society, Lindsay proclaimed that "Among all the marvelous applications of science to warfare that the great European war has produced—the gas shell, the 75-centimeter gun, the submarine, the Liberty motor—there is nothing more significant than the attempt to apply the principles of mutuality and insurance to lighten the burdens of war for our fighting men and their families and dependents." "As soon as America entered the European war," Lindsay recounted, "we realized that . . . the existing pension legislation . . . [was] as much out of date as the flint-lock musket."[79]

The WRIA was a piece of Progressive Era legislation par excellence. Supporters of it believed that the actuarial sciences (mathematical calculations of disability and risk) could solve the social problems brought about by war—disabled men, fatherless families, a beleaguered workforce, demanding veterans, a strained treasury—making their system of benefits more democratic, efficient, and just than Civil War pensions. The WRIA was, in short, a large-scale experiment in social engineering.[80] As such, it proved to be both legally and administratively more complicated than the Civil War pension ever was.

To get a sense of the complexity, one needs only to refer to the act itself.[81] Its purpose was threefold. First, it provided "allotments and allowances," money paid out to a soldier's dependents, covering their living expenses while he was away at war and in the event of his death. All US servicemen were compelled to contribute a minimum of $15 or at least one half of their monthly wage for the support of their dependents.[82] Canada was the only other belligerent country to mandate that soldiers made allotments to their dependents.[83] Second, the US government offered voluntary life and disability insurance to its soldiers at subsidized rates. For $.63 to $1.08, a soldier could get $1000 of coverage, a rate significantly lower than the $100 per $1000 rate offered by private insurance companies.[84] If all military soldiers took out insurance plans, the argument went, then there would be no future demands for service pensions.[85]

The third and final goal of the WRIA was to install a system of disability compensation based on a fixed, published rating schedule. The schedule was to "do away once and for all with the favoritism, arbitrariness, and individual pension bills that had marred the administration of the General Law Pension System, which applied statutory rates to certain specific disability but employed no rating schedule." The compensation feature of the WRIA was a direct extension of the workmen's compensation laws. G. F. Michelsbacker, an author of the disability rating schedule used by the California Industrial Accident Commission, drafted one of the original rating schedules used by the War Risk Insurance Bureau.[86]

With the passage of the WRIA, the meaning of what it meant to be both a US soldier and a male citizen changed dramatically. Like workmen's compensation, the WRIA reified assumptions concerning gender roles typical of certain Progressive reformers.[87] At its core, the WRIA promoted the belief that "husbands demonstrated their independence by providing a 'living wage' through paid work outside the home and by representing the family in the public sphere through voting, tax paying, and military service."[88] Wives, in turn, were supposed to remain at home and dedicate themselves to the unpaid labor of rearing future citizens.

Child labor laws, maternal health programs, mothers' pensions, and workmen's compensation all worked toward the end of protecting the American family by encouraging a safe and well-paying workplace for the father, allowing mother and child to remain at home. The WRIA was no different in this respect. Referring to the demands of war, Lindsay put it the following way: "under the conscription law the family is conscripted when the bread winner is taken away."[89]

Maternalists, such as Lathrop, who helped frame the legislation, supported the WRIA because it encouraged men to be out of the home, earning money. The insistence that all men (and only men)—whether veteran, disabled, immigrant, or poor—should be breadwinners explains why Progressive Era maternalists at once supported mothers' pensions (public stipends paid to single mothers) and opposed Civil War veterans pensions. Whereas the former encouraged mothers to remain at home and raise their own children (rather than relying on childcare outside of the home), the latter produced "home slackers," men who lived off the dole at a monetary rate barely large enough to support an entire family.[90] Mothers' pensions, it was argued, enhanced the civic and moral ordering of the nation and family because it allowed women to be the moral and social guardians of their children. Civil War pensions, on the other hand, threatened to disrupt the health of the nation by breaking down the family nucleus. To Progressives at the time who were concerned about rising divorce rates, declining reproduction rates among Anglo-Saxons, and the ill effects that frequent job turnover had on family ties, having a pension system that increased the numbers of non-wage-earning husbands only added injury to insult.

The goal of the WRIA was thus to encourage disabled soldiers coming back from the Great War to marry, have children, and become breadwinners again, working outside the home. Neither mere insurance nor disability compensation could assure such an outcome, as Mack and his fellow WRIA framers were well aware. Mack specifically worried that wounded soldiers would "come back in a more or less helpless condition," a physical state that would require wives and children to take care of the men of the household, rather than the other way around. More troubling to him, though, was the thought of all the unmarried conscripts who would come home disabled from war, with no wife and no job. "No barriers ought to be put in the way of marrying," Mack wrote, "and no barriers ought to be put in the way of good women in marrying [the veterans] because of the inability of these men to go out into the world to make the sort of living that other people . . . would have."[91]

In the hopes of making the disabled soldier's return barrier-free, Mack, Lathrop, and the other framers of the WRIA mandated medical care and rehabilitation services for all returning disabled soldiers, writing these services into law.[92] From Mack's perspective the analogy was clear: just as rehabilitation could make a juvenile criminal into a law-abiding citizen, so too could it turn a disabled soldier into an able-bodied wage earner. Whereas "the man who lost a leg at Gettysburg was cobbled up as well as the surgery of 1863 could do it, given a pension, and turned loose," as one commentator put it, the new system of medical aftercare would cure the amputee of his dependency by putting him on his "feet" again.[93]

The unwavering belief that the medical sciences could make disabled soldiers into productive wage-earning citizens set the WRIA well apart from the Civil War pensions system. The General Law, with all its permutations throughout the late nineteenth century, did not place the same emphasis on a veteran's bread-winning capacity as the WRIA did. Unlike the WRIA, the Pension Bureau cut checks based on severity of disability, not on number of dependents, and not on wage-earning capacity. Nor did the Civil War pension system institutionalize medical services the way the WRIA did. To be sure, medical doctors had long been involved in the disbursement process of veterans' benefits; as already noted, the Civil War pension system required that all claimants secure a physician's medical diagnosis of disability before an application for pension payment could proceed. Yet the WRIA made medicine integral to the entire benefits process, turning it into a part of the compensation package itself.

The revolutionary significance of the WRIA's mandate of medical care did not escape contemporaries who heralded it as the "most important contribution" to the evolution of compensation laws. Up to that point, workmen's compensation laws had not succeeded in securing medical care or rehabilitation for laborers, despite the fact that early studies revealed that without medical attention, skilled laborers often ended up performing unskilled work after a disabling accident.[94] Commenting on the noteworthiness of the WRIA's coverage of unlimited medical care, University of Chicago economist and later Democratic senator from Illinois Paul H. Douglas wrote in 1917: "the provision for the rehabilitation and re-education of the disabled. . . . deals with a vital problem that has been almost universally neglected by compensation legislation. The disabled person needs, not only a money grant, but also training, so that his disability will be the slightest possible hindrance to his re-entering industry."[95]

Put another way, Progressive reformers believed that rehabilitative medicine could solve the problem of the pensioner. "One of the most useful and necessary duties of this department," wrote one War Risk Insurance Bureau chief, "will be to prescribe and furnish medical and surgical treatment in order that *disabilities may be reduced or caused to disappear entirely*" (emphasis added).[96] This kind of confidence in the powers of the new medical sciences was common among Progressive reformers, especially those involved in the New Public Health, who used the recent bacteriological sciences to track down germs and their carriers in order to prevent the spread of contagious disease. The proposal that medicine could cure disability, however, was something unique, for neither the bacteriological sciences nor the cutting-edge discoveries in pharmacopoeia could help restore the mutilated bodies coming back from the war to the point of full recovery. The faith that disabilities could be cured came from abroad, from Continental Europe and Canada, where war-torn soldiers were made whole again through programs in medical rehabilitation (a subject taken up in the next chapter).[97] The WRIA framers heard of these success stories and believed that rehabilitative medicine should be made into law.

The United States took the institutionalization of rehabilitation one step further than its wartime allies, however. Unlike Europe and Canada, where rehabilitation was largely a voluntary component of a disabled veteran's benefits package, the United States compelled disabled servicemen to undergo long-term medical treatment. Framers of the WRIA incentivized medical care and rehabilitation through law by suspending compensation payments to any disabled soldier who refused to undergo hospital care.[98] Here again, framers of the WRIA were reacting to the country's history with the Civil War pension system. In their minds, if rehabilitation was compulsory, then more men could be restored to the workplace, thereby reducing the costs of federal disbursements.

*   *   *

While many early twentieth-century Americans revered the Civil War veteran, celebrating his heroism through patriotic parades and continued support of federal benefits, certain Progressive Era reformers found him to be the enemy within. To these Progressives, the Civil War pensioner threatened the moral fabric of the country. He stood in the way of putting the War of the Rebellion in the past. He encouraged a political system based on favoritism and patronage rather than on neutrality and fairness. He made money by filing claims with the Pension Bureau, rather than by the sweat of his own brow. He lived in federally funded soldier's homes, rather than in his own private dwelling with a wife and children. He greedily took and never gave anything back in return.

The War Risk Insurance Act of 1917 promised to put an end to all of this. A crucial—and indeed revolutionary—component of the WRIA was the institution of a large-scale system of rehabilitative medical care. The Progressives who drafted the WRIA pinned their hopes on the medical sciences of the day, believing that rehabilitation was a surefire way to get the nation to break its so-called "national pension habit." Disabled soldiers coming home from the Great War would be put to work in medically based curative workshops, where they would learn how to use their maimed bodies in order to become laborers once again.

TWO

Reconstructing Disabled Soldiers

The framers of the War Risk Insurance Act saw medical re-habilitation as a means of conservation, a way to preserve the nation's economic and human resources. Human ener-gies, depleted by war, had to be replenished. Human bod-ies, maimed in battle, needed to be rebuilt. Time and again, those in favor of rehabilitation insisted that the principles of scientific management would "trump the lingering irratio-nality of industrial war, and manifest itself in reconstructed veterans' bodies once they matched the appropriate labor venue to the appropriately remade man." Rehabilitation promised to conserve labor power and thus enhance na-tional efficiency in times of war and peace.[1]

Orthopedic surgeons became the experts of choice to ac-complish the goal of reconstructing US disabled soldiers from the Great War. That they would become the overseers of reforming the nation's policy toward its disabled soldiers was not preordained. By the time of America's involvement in the First World War, there were approximately one hun-dred self-identified orthopedic surgeons (all male) practic-ing in the United States, a relatively small number compared with the better-established (and better-known) medical specialties of ophthalmology, obstetrics, and gynecology.[2] Of the few physician-surgeons who practiced orthopedics, most lived and worked in one of three cities: Boston, New York, or Philadelphia. A few solitary practitioners could be found in Baltimore, St. Louis, Minneapolis, Pittsburgh, and

Chicago. In short, orthopedic surgery was a relatively esoteric specialty, only to be found in the nation's population centers.

What's more, a significant part of orthopedic practice concerned children who suffered from chronic diseases and congenital deformities. Looking at the records of patients treated at the Philadelphia Orthopedic Hospital during the early twentieth century, for example, one finds that infants and toddlers populated the wards and waiting rooms, their parents looking for remedies to manage painful (and sometimes life-threatening) deformities and diseases such as Pott's disease (tuberculosis of the spine), dislocated hips, rickets, clubfoot, and infantile paralysis.[3] Some of the earliest hospitals devoted to orthopedic care included the State Hospital for the Indigent, the Crippled and Deformed Children in St. Paul, Minnesota, as well as the children's hospitals of Boston, St. Louis, and Pittsburgh.[4]

How and why a small group of medical specialists who largely focused on chronic pediatric conditions became responsible for the monumental task of establishing institutions of medical rehabilitation abroad and at home for thousands of acutely disabled American soldiers is the subject of this chapter. Orthopedic surgery's strong international ties to fellow orthopedists practicing on the other side of the Atlantic is one explanation for the profession's wartime ascendancy. With the European continent engulfed in a total war that threatened to wipe out an entire generation of young adult men, belligerent countries began instituting programs in orthopedic rehabilitation during the initial years of the war with the hopes that maimed men could be reused in battle, sent to the front line despite their disabilities.[5] American orthopedic surgeons became engaged with the Great War—intellectually and physically assisting their foreign colleagues—well before the United States officially aligned with the Allied Forces in the spring of 1917. American orthopedic surgeons not only felt an alliance to their country, but also to their profession, which, since the founding of the American Orthopedic Association (AOA) in 1887, grew beyond national borders. During the first decade of the AOA's existence, its membership roster included famed orthopedists from other countries, such as William John Little, Hugh Owen Thomas, and Sir Robert Jones, all from Great Britain; Adolf Lorenz, from Austria; Julius Dollinger, from Hungary; Fitz Lange, from Germany; and, later on, Vittorio Putti, from Italy.[6]

These trans-Atlantic relationships proved to be essential in granting US orthopedic surgeons legitimacy as experts on the rehabilitation of disabled soldiers during and after the First World War.[7] Less than four months after the United States declared war on Germany, Surgeon General William C. Gorgas established the Division of Orthopedic Surgery, separate from both the Division of General Surgery and the Division of Medicine.[8] Prior to the Great War, orthopedic surgery had not been a recognized military medical specialty in the United States. During both the Civil War and the Spanish-American War, general surgeons took care of all types of battle wounds and were responsible for everything from amputating limbs and removing bullets lodged in flesh to bandaging fractured bones.[9] Following the example set in motion by both the Allied and Central Powers, the US Army of the First World War gave orthopedic surgeons a level of administrative legitimacy and autonomy they had never before experienced.[10]

But the war was not the only, or even the primary, reason why orthopedic surgery came to be seen as the profession of choice in instituting rehabilitation programs for disabled soldiers. Through their prewar work with disabled children, orthopedic surgeons shared a common vision with Progressive reformers who wanted to dismantle the Civil War pension system. Just as pension reformers complained about one-armed and one-legged Civil War veterans earning a living by begging, so too did orthopedic surgeons criticize their crippled patients who, after receiving medical attention, took to the streets, looking for alms.[11] Consider the words of University of Pennsylvania orthopedic surgeon Gwilym G. Davis in 1914: "A cripple is a menace both to himself and the community and is apt to become a burden on his relatives, his friends and the public. The aim then is to improve his physical condition and character as to make him . . . self-supporting, self-respecting, self-reliant and able and willing to take and perform his part in the communal life."[12] With the knowledge that a strong correlation existed between disability and poverty, many orthopedic surgeons became deeply involved and politically engaged in the late nineteenth- and early twentieth-century Progressive movements toward welfare reform—a reform that differentiated the worthy from the unworthy poor by looking at productive capacity rather than at a person's sins and virtues.[13]

Long before Civil War pension reform came about, orthopedic surgeons had been engaged in establishing industrial training schools (primarily for children) in order to teach their patients trade work, making them self-supporting. Here the Protestant work ethic merged with medical therapeutics.[14] Disability could be cured, orthopedic surgeons believed, only when the patient became employable. Orthopedic surgeons spoke in terms of the "mental and moral training" of patients, teaching them to overcome their natural tendency toward idleness.[15] Such training also required public outreach, educating parents, schoolteachers, private philanthropies, and the general public to view the disabled not as charity cases who should receive free money and so-called coddling, but rather as potentially productive citizens who, under proper medical supervision and care, could become independent and self-sufficient.

Orthopedic surgeons gained significant stature among US reformers and the military during the First World War because they defined themselves as "social guardians" of the nation, seeing their work as something that went beyond medical practice.[16] While significant changes were taking place in bacteriology (i.e., the germ theory of disease) and in general surgery (i.e., aseptic techniques allowing surgeons to operate with relatively low mortality rates) during the late nineteenth century, the practice and theories of orthopedic surgery remained fairly static.[17] Aside from a handful of new (and minor) operative techniques developed over the course of the nineteenth century, the orthopedic surgeon of the 1910s looked very much like his counterpart practicing in the 1860s: men of both generations relied on trusses, braces, traction, and other mechanical devices to straighten crooked backs and limbs and to bring relief to painful joints.[18] The novelty that orthopedic surgery offered came in the form of medical welfarism, a new kind of "scientific charity" in which medical theory validated the felt social need to make the disabled useful through the production of capital.[19] Their interest in the science of charity, rather than the science of medicine, catapulted orthopedic surgeons to the forefront of bringing about the reform that Progressives involved with the War Risk Insurance Act had been looking for.

Orthopedics and Child-Saving

When the New York Hospital for the Ruptured and Crippled first opened on Lexington Avenue in 1870, the US federal government estimated the national infant mortality rate to be 175.5 deaths for every one thousand children.[20] Abraham Jacobi, known as America's father of pediatrics, would have given a higher number. Working as an attending physician in New York's foundling homes and Children's Hospital during the last third of the nineteenth century, Jacobi estimated that as many as 75 percent of the city's children were dying from malnutrition, infectious disease, and psychological neglect.[21] The threats to a child's health were many in the late nineteenth century, ranging from contaminated water and milk to acute infections such as measles, diphtheria, mumps, whooping cough, scarlet and rheumatic fever.[22]

Dr. James Knight, founder of the Hospital for the Ruptured and Crippled, was more interested in managing the physical aftermath of disease and malnutrition than in preventing such maladies from occurring in the first place. Knight received his medical degree in 1832 from the Washington Medical College in Baltimore. For the remainder of the decade, he practiced as a family physician and when he was not seeing private patients, he volunteered his time at the Baltimore General Dispensary. On the advice of Dr. Valentine Mott, who told him that more physicians were needed in orthopedics, Knight moved to New York City in 1840 to participate in orthopedic clinics held at the University of New York medical department.[23]

The orthopedic clinic in New York catered to a specialized group of patients who suffered from chronic conditions related to the musculoskeletal system. The work of an orthopedist, as one practitioner would put it, required treating the type of patient who "neither dies nor gets well, but is disabled more or less permanently."[24] Knight considered himself an expert in the "aberrations of the human form."[25] Orthopedic deformities occurred either congenitally or as a result of childhood diseases. Common deformities from birth included clubfoot, dislocated hips, and cerebral paralysis.[26] Acquired deformities arose later in childhood, usually as sequelae to either nutritional or infectious diseases that remained incurable before the development and widespread use of synthetic vitamins during the first third of the twentieth century and antibiotics during the 1940s. Rickets, a vitamin D deficiency that causes softening of the bones, often affected the lower limbs, leaving children with either bowed legs (genu varum) or knocked-knees (genu valgum).[27] Tuberculosis of the spine, known more commonly at the time as Pott's disease, resulted in humpbacks, permanent and obvious curvatures jutting out from the backbone.[28]

Orthopedists such as Knight would treat these afflictions mostly in the postacute stage, using bracing, traction-and-pulley tables, and exercise with the goal of realigning deformed bodies. While the pursuit of such realignments arose from aesthetic and moral convictions about the virtues of having an upright body (an argument associated with such writers as Michel Foucault), this was not the only motivating force behind the practice of orthopedics.[29] Many patients came to Knight's orthopedic clinic looking for pain relief. Others came believing that bracing offered a better alternative to the more radical (and risky) operative procedures that required the knife. A good example to demonstrate this point is Knight's treatment of inguinal (groin) hernias. In contrast to some surgeons who would sever the hernia and cauterize the groin—a life-threatening procedure before antiseptics—Knight fabricated wearable braces and trusses that would push the protruding contents of the abdominal cavity back into place.[30] After his move to New York, Knight became known city-wide for his "exceptional mechanical ability [in] designing trusses and abdominal supports."[31]

Knight treated both adults and children, but in his practice, as in most orthopedic clinics, the two populations remained segregated. By and large, adults received orthopedic care in outpatient dispensaries, whereas children became inpatients at hospitals and sanatoriums where they could be continuously monitored by the medical staff. In the days before health care insurance and workplace benefits, both products of the twentieth century, most working-class adults did not have the money to pay the boarding costs of an extended hospital stay, nor could they afford to miss a day's, let alone a week's, worth of wages.[32] Still, as a specialty that aimed to alleviate the aches, pains, and injuries of the musculo-skeletal system, orthopedics was, in many ways, medicine for the laboring class. Knight understood this fact, making the original charter of the Hospital for the Ruptured and Crippled explicit in its goal of serving the poor who could not afford orthopedic devices or treatment. The institution was to "supply skillfully constructed surgico-mechanical appliances, . . . trusses, . . . bandages, lace stockings and other suitable apparatus for the relief and cure of cripples, both adult and children." "And so far as possible," the charter concluded, "make these benefits available to the poorest in the community."[33]

From the hospital's beginning, children with orthopedic conditions received special attention. With the financial help of the New York Association for Improving the Condition of the Poor and the New York Society for the Relief of the Ruptured and Crippled, Knight established the original 28-bed hospital in his own Manhattan home in 1863 for the purpose of treating children exclusively. Here he enforced a daily exercise regime and provided educational instruction of "both the religious and secular" kind. Convinced that bodily excitation through exercise cured disease processes, Knight vigilantly drilled his patients, keeping them in motion.[34] "No child able to hold up its [sic] head," he wrote, "is ever kept in bed during the day." Those who were able to walk, but who needed assistance, would "push a chair before them," and those who were in too much pain to ambulate would be "supplied with rolling chairs."[35]

Knight's hospital was not the only one of its kind. Several other institutions devoted to the care of crippled children existed in New York City. In 1866, Dr. Charles Taylor opened the New York Orthopedic Dispensary, with a private sanatorium for children attached. In 1861 Dr. Lewis

Sayre opened an orthopedic dispensary at the Bellevue Hospital. Similar institutions existed in Boston and Philadelphia. Dr. John Ball Brown, a consulting surgeon at the Massachusetts General Hospital, founded the Boston Orthopedic Institute for the Cure of the Deformities of the Human Frame in 1833. Three decades later, his son, Buckminster Brown, helped to establish not only a small private hospital, the House of the Good Samaritan, which specialized in orthopedics, but also the Boston Children's Hospital in 1869. Likewise, the Philadelphia Orthopedic Hospital was incorporated in October 1867.[36]

Although these physicians and their institutions treated a mixture of children and adults, young patients held a unique place in the practice and history of orthopedics. French physician Nicolas Andry coined the term *orthopaedia* in 1741, combining the Greek roots *orthos*, "straight," and *pais*, "a child." He wrote his two-volume book *L'orthopédie* to serve more as a child-rearing guide than medical manual.[37] Keeping company with other French Enlightenment thinkers, such as Claude Helvétius, Andry believed that proper education was necessary to bring about improved health and hygiene in society.[38] Children became the focus of this campaign, for Andry believed that they were closer to nature, innocents who had not yet been corrupted by the conditions of the material world.[39] He blamed many childhood deformities on improper handling and rearing—a charge that fell squarely on the shoulders of mothers and wet-nurses.[40] In his book, he instructed caregivers on proper swaddling techniques so as to avoid infant deformities, on balancing exercises to keep young backs erect, and on daily grooming techniques to maintain shapely nails, eyebrows, and ears.

The emphasis on the orthopedic treatment of children, rather than adults, increased in the early nineteenth century when subcutaneous tenotomies became a safe and accepted practice for treating congenital clubfoot. A newly designed knife, the tenotome, had a blade thin enough to allow for the dividing of tendons and fascia without causing permanent trauma to the other structures of the ankle and foot. With this instrument, orthopedic surgeons could safely divide the Achilles cord, making it lax, so that turned-in feet could be made perpendicular with the shin, mimicking normal anatomical development. The technique was perfected and subsequently popularized by Hanover physician Georg Friedrich Louis Stromeyer in 1831.[41] Many of America's first orthopedic surgeons became adherents to Stromeyer's techniques. William Detmold, an émigré from Germany, opened a clinic for crippled children in New York City in 1841 (later connected with Bellevue Hospital) and performed

167 tenotomies in the first three years.[42] John Ball Brown expanded the use of tenotomies to correct torticollis (wry neck) and scoliosis. Reflecting the importance of tenotomies and the rise in status and practice of orthopedic surgery in a journal article in 1844, Brown wrote: "The Art of Orthopedy is not of recent origin, it was practiced a hundred years ago [by Andry]—but the discovery (of recent date) that the tendon could be divided with impunity gave new life to the most useful, but which had become an obsolete, art."[43]

The emphasis on childhood diseases and deformities in nineteenth-century orthopedics, however, cannot be explained by etymology and technology alone. More than a handful of orthopedic surgeons came to the specialty as a result of their own personal experiences with the disabling (and sometimes fatal) effects of diseased bones. John Ball Brown, originally trained as a general surgeon, became interested in orthopedic surgery in the early 1830s as a concerned father: in 1833, his son Buckminster, at the age of fourteen, had been diagnosed with Pott's disease. Similarly, DeForest Willard, one of Philadelphia's first self-proclaimed orthopedic surgeons, committed himself to a medical education out of a desire to correct his own disabilities. When he was an infant, Willard suffered a severe attack of poliomyelitis, leaving him with what is described in the historical record as "one partially paralyzed leg and a clubfoot deformity," as well as an undetermined visual impairment, all of which persisted into adulthood.[44] He originally wanted to attend Yale College and had every reason to expect to be accepted; not only did he come from a privileged family in Newington, Connecticut, but he also passed the Yale entrance exam immediately after graduating from Hartford High School. Despite his proven intellect and pedigree, Yale rejected Willard because of his physical disabilities.[45] As a result, in 1863, Willard traveled to Philadelphia to study medicine at Jefferson College. After completing his first year at Jefferson, he allowed University of Pennsylvania surgeon, D. Hayes Agnew, to perform a tenotomy on his club foot, a procedure that proved to be "markedly successful," even though one of Willard's legs remained shorter than the other, requiring him to walk with the assistance of a cane. From that point forward, Willard devoted his career to orthopedic surgery, completing his last year of medical school under Agnew's tutelage.[46]

In addition to those who came to orthopedic surgery through their own childhood experiences with disability, others came to it out of frustration with what little nineteenth-century medicine had to offer in terms of effective treatments. Aside from tenotomies for clubfoot, no quick fix existed for most of the conditions that passed through an orthopedic surgeon's office. No one story encapsulates the despondency that orthopedists sometimes felt better than that of Dr. Taylor's rendition of the "Gilbert girl" case. In the fall of 1858, Taylor examined a six-year-old girl who exhibited a small "knuckle" in the back and a "heavy resting against the mother's knee when she approached her." Although Mrs. Gilbert felt that her child was "as well as ever, as well as any child," Taylor told her that her daughter had a "destructive disease in her spinal column"; in other words, early stage Pott's disease. When the mother asked what she should expect, Taylor explained the degenerative nature of the disease, telling her that the disabling effects would grow greater and greater as time went on. The mother replied:

> Do you mean to say that my little girl here, who has been always apparently a healthy child, and now appears to be so, who is still so plump and straight, must go down before my eyes, must suffer, may be confined to bed with paralysis, with discharging abscesses [a common sequela of tuberculosis of the spine], down through a series of years, becoming deformed and dwarfed and emaciated, distorted and disfigured, and there is nothing which can be done to prevent, or even to mitigate, such a terrible calamity?[47]

Before antibiotics, the cure for tuberculosis of the bone existed in sanatoriums, in solariums, and in the orthopedists' appliance shop—this last a place where Taylor thrived, using skills picked up from a local blacksmith who taught him how to mold metal and leather around a patient's body in order to forestall crippling deformity.[48] After seeing more and more cases similar to the Gilbert girl, Taylor invented the "spinal assistant," a torso brace that prevented forward bending but allowed backward bending—a design intended to relieve pressure on the frontmost, weight-bearing part of the vertebrae, where the disease first set in. By Taylor's account, once he began applying the brace, he saw immediate results, even cures.[49]

Virtually every success story that Taylor told was of a child patient, not an adult. Part of the reason stems from the fact that at the time tuberculosis of the spine was believed to occur more frequently in children than in adults. Virgil Gibney, the successor to Knight at the Hospital for the Ruptured and Crippled, estimated that 87 percent of Pott's disease occurred in patients younger than fourteen years of age. Taylor concurred, contending that out of the 375 patients he had seen by 1881, 226 were under five years old, 68 were between the ages of five and ten, and the remaining 24 between ten and fifteen years of age.[50] Whether or not these numbers reflected actual disease rates is difficult to determine, for

1   An example of Pott's disease affecting the thoracic spine. James K. Young, *A Manual and Atlas of Orthopedic Surgery* (Philadelphia: Blakiston's Son & Co., 1905), 225.

adult patients with Pott's disease might not have sought the assistance of orthopedists and institutions devoted to the care of children.

Regardless of disease occurrence and etiology, orthopedists preferred to treat younger patients because they believed that their bodies (and minds) were more malleable and manageable than adults. According to medical doctrine at the time, treatment for Pott's disease could take anywhere from fifteen months to several years. What the patient was supposed to do during his or her convalescence was a point of contention among practicing orthopedic surgeons. Some believed that complete bed rest would bring about the best results. Orthopedist A. J. Steele, of St. Louis, recommended complete recumbence on a steel frame, with fasteners to strap the patient in place, keeping the child as motionless as possible. Steele claimed to have had patients "lying in the recumbent position for eleven or twelve months whose general health continued

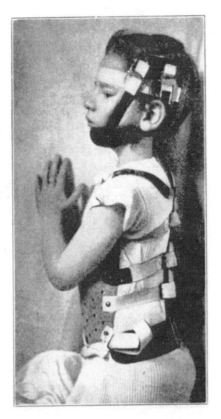

2   A version of Taylor's "spinal assistant," bracing used to treat tuberculosis of the spine. James
    K. Young, *A Manual and Atlas of Orthopedic Surgery* (Philadelphia: Blakiston's Son & Co.,
    1905), 258.

excellent."[51] "I do not believe it is necessary," wrote Steele, "that young
children should be allowed to walk and run about as far as their general
health is concerned."[52] Another proponent of complete rest was Robert
Lovett, of the Boston Children's Hospital. Convinced of the imperfection
of bracing—that no mechanism adequately relieved pressure from the
diseased vertebrae—he told his fellow orthopedists at the annual AOA
meeting in 1896 that in the acute stage of Pott's disease "it is incumbent
upon the surgeon either to insist upon . . . recumbency or to transfer the
responsibility [of treatment] . . . to the parents."[53] In other words, Lovett
refused to treat patients who refused rest.

   On the other side of the debate stood those orthopedic surgeons who
contended that complete rest was not only impossible to achieve but
also worsened the patient's health. Gibney believed that immobility

made diseases fester. In direct response to Lovett's 1896 proclamation, he retorted: "You cannot keep them [the patients] in bed."[54] Reginald H. Sayre, another student of Knight's, responded similarly, arguing that complete rest was impossible to achieve, even in recumbency. Arguing from anatomy, Sayre pointed out that patients with Pott's disease of the thoracic spine would never experience complete rest since respiration caused "constant trauma" to the vertebra attached to the ribs.[55] A colleague of Gibney's, A. M. Phelps concluded the discussion of Lovett's paper, arguing "the plan of putting patients in bed is an old one."[56] Phelps maintained that complete rest would make the problem of psoas abscess (pus-filled sac, bulging from the inner thigh resulting from the descent of the tuberculosis bacteria downward toward the groin) more pronounced. In severe enough cases, orthopedists performed needle aspirations to drain the abscess. If mishandled, abscesses could lead to septicemia and ultimately death. Phelps contended that if a patient remained reclined in bed, the abscess would "burrow and discharge through the bodies of the vertebrae." But if the patient remained ambulatory, "the pus [would] find [an] exit from the body."[57] In lieu of complete bed rest, Gibney and Phelps advocated aggressive bracing, immobilizing the back, but in a way that still allowed the patient to walk about.

No matter the treatment, both bed rest and bracing required a degree of control over the patient's body and everyday life that could, orthopedists believed, only be effectively exerted over children.[58] Adult back deformities could be braced and laced up in the clinic, but the doctor had little control once the patient left his office. Time and again, orthopedists complained about patients who would tinker with their braces at home, adjusting straps and harnesses to make them more comfortable to wear. But comfort, in the orthopedists' mind, often worked against the cure: forcing a curved spine to remain erect required a degree of pain. Orthopedists saw similar noncompliance in brace wear with their child patients, except in these cases the mothers were to blame. Engaging in the same gender-laden scapegoating that Andry had 150 years earlier, Phelps complained of mothers and wet nurses who "monkied around" with braces when not under the orthopedist's eye. "The [child] complains of pain somewhere," Phelps explained, "and then the mother thinks she must take it off . . . or loosen it in some way, and the result is in these cases the kyphosis [back curvature] increases." So although Phelps opposed bed rest for his patients with Pott's disease, he nevertheless supported constant supervision of the patient, the kind that could only happen if the child were institutionalized.[59]

Adult men were assumed to be breadwinners of the family and, thus, unable to undertake prolonged convalescence. "I remember one [adult] patient," Sayre told his fellow orthopedists in 1896, "who had been advised that he could not recover without spending a couple years in bed and the question presented itself . . . as to how he was going to support his family while lying in bed for years waiting for his spine to recover."[60] Women, too, were expected to work, either raising their children or finding outside employment in domestic jobs. Only the wealthiest patients—those who did not need a weekly wage and who could hire help to manage the home and raise the children—could afford bed rest.[61] Looking at the hospital ledgers and published records of most orthopedic hospitals and clinics of the late nineteenth century, patients with this kind of wealth were rarely seen.

The emphasis on childhood, however, was not simply internal to the practice of orthopedics itself. The late nineteenth century, after all, was a time when American social reformers of all types rallied around "child-saving." "Throughout the country," historian Michael Katz writes, "by the 1890s children had captured the energy and attention of social reformers with an intensity never matched in other periods of American history." Many anxieties fueled the child-saving movement, from immigration and urban growth to political radicalism and the economic downturn of the 1893 depression. As Katz puts it, "child-saving was one key strategy for stemming the slide of the poor in great cities into savagery, hostility and socialism."[62]

The crippled child played an essential part in this vision of reform. Douglas McMurtrie, a well-known Progressive Era bibliographer, print historian, and advocate for the disabled credited orthopedic surgeons with being the first group of professionals to "realize . . . the special needs of cripples," while "regular national organizations engaged in philanthropic [work had] given no attention to cripples."[63] A separate 1914 study funded by the Russell Sage Foundation came to a similar conclusion, stating that "the credit for [the creation of] institutions for crippled children is due chiefly to the unselfish devotion of the orthopedic surgeons who lead the movement."[64] McMurtrie heralded the efforts of orthopedic surgeons such as Knight, who had convinced New York philanthropists to pay the cost of constructing and maintaining the Hospital for the Ruptured and Crippled. The creation of hospitals and homes devoted to the care of crippled children ballooned during the late nineteenth century, resulting in approximately 70 such institutions by the first decade of the twentieth century.[65]

There were many ways in which saving the disabled child differed from the more general reform tactics targeted at able-bodied children.

Most strikingly, whereas the general trend of saving able-bodied children was toward deinstitutionalization—getting needy children out of poorhouses and into foster homes—the opposite held true for disabled children.[66] As Edith Reeves of the Russell Sage Foundation pointed out: "That the family home is the best place for well children is now generally recognized. But crippled children are conceded to be a special class, requiring in many cases surgical operations and in most cases very close physical supervision for months, often years." In addition, reformers interested in saving the disabled child believed that "crippled children in family homes are not usually treated as other members of the family are; that they are petted, or in some cases neglected and despised, if they live at home." The latter danger only increased, it was argued, if the child was sent to a foster-care home where reports of abuse of the disabled ran rampant. Using the logic that disabled children could "never know . . . a normal family atmosphere," orthopedic surgeons and their reformer compatriots argued that hospital-like institutions were the safest and healthiest places for such children to thrive.[67]

The Widener Memorial School for Crippled Children in Philadelphia, established in 1906, became a model institution for reformers both inside and outside of the medical profession. Before the turn of the twentieth century, Dr. Willard approached millionaire philanthropist Peter A. B. Widener to fund the construction and maintenance of a long-term care facility for children with orthopedic disabilities. At the time, Willard had just succeeded in opening a children's orthopedic ward with a fully equipped gymnasium at the University of Pennsylvania hospital.[68] But in Willard's mind, the facilities were too small and the average duration of stays too short. Children with orthopedic disabilities, he believed, needed years of care. Most important, he contended, the medical profession needed a place "to educate and train the crippled child . . . [toward] self-support."[69]

Convinced of the worthiness of Willard's plan, Widener purchased thirty acres of land at the corner of North Broad and Olney Avenue and erected a main hospital with twenty colonial-style buildings surrounding it. The campus, trimmed with ornamental landscaping, also included a 10-acre vegetable garden and multiple playgrounds, all at the cost of over one million dollars. Teachers hired at the Widener followed the curriculum and examination standards set by the public schools of Philadelphia. In addition to classroom time, patients of the Widener would take "hand work and vocational training" classes, including sewing, cooking, basketry, stenciling, caning, woodwork, carpentry, cobbling, farming, and engraving. Admission was limited to children between the ages of four and ten, and patients would remain at the Widener until "after they reach[ed] the age of twenty-one, perhaps at eighteen, if ready to begin wage-earning."[70]

In an era when reformers passed legislation to protect children against workplace abuse through labor laws, the insistence that patients at the Widener engage in vocational training and labor seems anathema.[71] Reformers at the time saw no paradox. Whereas children working in industrial mines and factories were subject to unsafe working conditions and slave wages, the patients at the Widener were being groomed for skilled labor; they also enjoyed a life free from destitution and a controlled workspace that functioned as a quasi-classroom. Vocational training was geared toward making upright citizens for the future, to instill the skills as well as the time and work discipline necessary to give disabled children an edge in the workforce once they became adults.[72] Programs of vocational training were not unique to Widener. Instilling the work ethic was a goal of many of America's earliest penitentiaries, reform schools, and mental hospitals.[73] In addition to these early attempts, the public school movement of the late nineteenth and early twentieth centuries embraced vocationalism as a key component to universal education. Many Progressive Era educational reformers believed that the central task of the school "was to integrate youth into the occupational structure."[74] The difference with the Widener's vocational training was that it was done in the name of physical medicine. Again, labor was seen as a way to cure disability, making the crippled into able-bodied, productive citizens.[75]

As a result, successful orthopedic treatment increasingly became measured by the number of patients who entered the workforce, for meeting such a goal, it was argued, meant that the disabled were no longer handicapped by their physical condition. Securing employment, orthopedists believed, put the disabled on par with able-bodied citizens. Harvard orthopedic surgeon Dr. Edward Bradford, who founded the Industrial School for Crippled and Deformed Children in Boston in 1894, published charts and graphs tracking the employment of his patients as a way to legitimize the existence of the institution and attract potential patients and donors. The graduating class of 1910 had great success on the job market, with one patient who had an amputated leg earning $20 a week (what would amount to approximately $450 per week today) as a lunch cart tender. Another patient with Pott's disease landed a job as a "sales girl" for the weekly pay of $10.[76]

With employment as the gauge for medical efficacy, unemployment—or worse, begging on the streets—was taken to be a sign of therapeutic failure. During his president's address to the AOA in 1914, Davis recounted the following example, albeit of an adult patient, of how orthopedics could fall short: "On one occasion I gathered together enough money to provide a one-legged man with an artificial leg. When however he found that if he used the leg it would interfere with his begging and posing as a poor, helpless cripple then he would not take the artificial leg as a gift." A decade earlier, Gibney related a similar experience to his fellow orthopedists, telling them that he had "seen former inmates of the Hospital for the Ruptured and Crippled begging on the street corners." Orthopedists, however, rarely blamed themselves for such failures. At times, they blamed the patient, deeming the person noncompliant or of "bad moral character." At other times, as demonstrated above, female caregivers were held responsible. But by the late nineteenth century, orthopedic surgeons increasingly condemned the public and the US government for not adequately supporting the work of orthopedics and the building of institutions, such as the Widener, where the disabled would be provided with a "proper mental and moral" training and thus be dissuaded from the "irresistible temptation of mendicancy."[77] "The public itself," wrote Davis on the eve of the First World War, "is to a great extent ignorant and sometimes indifferent as to both what should be done for the cripple and how it should be done." Echoing the concerns of Civil War pension reformers, Davis concluded that the beggars who populated the nation's city streets were "glaring examples of the incompetence and indifference of [the US] government and [its] people."[78]

At a time when social Darwinism affected social policy and reform in the United States, orthopedic hospital-schools attracted support among middle-class professionals because such institutions appeared to make it possible for disabled children to keep pace—to compete in the great struggle of life—with their able-bodied counterparts.[79] Before institutions such as the Widener opened, most disabled children were precluded from attending public school, where attendance, punctuality, and long hours of sitting in desks were required. Hospital-schools, by contrast, offered accommodations for children whose bodies could not comply to such strict, disciplinary measures. At Bradford's school in Massachusetts, for example, children who were unable to remain in their seats for prolonged periods of time "were sent to rest," assisted by a school nurse who would "loosen their apparatus" before the children lay down. The nurse would then reapply and readjust braces before students returned to their desks.[80] Nebraska orthopedic surgeon Winnett H. Orr pointed out the hypocrisy of reformers who instituted public schools for all children in order to stave off crime and poverty but did not allow equal opportunity for disabled children who, in his mind, were more likely to end up in poverty than their able-bodied counterparts. "The crippled and deformed," Orr pronounced at a 1909 New York symposium about the place of the disabled in society, "have almost no opportunities [for education] at all." "It would seem" he concluded, "as if the state were encouraging complete dependence of the physically handicapped."[81]

To correct this oversight among child-saving reformers, orthopedic surgeons joined with wealthy philanthropists to uplift not only the disabled child but also US society as a whole. In the years from 1891 to 1912, thirty-two institutions devoted to the medical care and vocational training of the crippled child were established. Some were state institutions, such as Minnesota's State Hospital for Indigent Crippled Children and Nebraska's Orthopedic Hospital (with Orr as its director), and others were private.[82] For the most part, orthopedic surgeons were in charge of these institutions. As such, they considered themselves to be a part of an important social reform movement—a movement intended to conserve both physically deformed bodies and the structures and functions of a wage-labor economy. Orr once referred to orthopedics and its institutions as forms of "prophylactic philanthropy." "It is easier and cheaper," Orr wrote, "and far more satisfactory to put these patients on their feet socially as well as physically than to be compelled to provide maintenance and special care for them during the many years of dependency which must inevitably come if such dependency has not been provided against."[83] It was this vision of orthopedics—to guard against the disabled becoming permanently dependent on state funds—that made them natural allies with the reformers who wanted to dismantle the Civil War pension system before the War Risk Insurance Act even came to pass.

Internationalism, Adult Orthopedics, and the First World War

American orthopedists had always treated adults, even though their primary institutional affiliations were with those devoted to children. One reason why the first generation of orthopedists in America could not devote their practice solely to treating adults was that there was little financial incentive to do so. The cause of saving the crippled adult would not begin until the passage of workman's compensation laws, and even then, medical care was rarely, if ever, guaranteed.[84] Therefore, few institutions catered exclusively to the crippled adult. In an 1896 symposium concerning the care of children with Pott's disease, Gibney, who treated adults in the outpatient dispensary connected to the Hospital for the Ruptured and Crippled, maintained that orthopedic surgeons "were not quite equal to the management of adult cases of Pott's disease, especially when complicated by abscess." He lamented the fact that because adult patients could not be admitted to the children's hospital, they had to be referred to local general hospitals, many of which would not accept such patients because it was believed that they would inevitably die from septicemia. Gibney defied such assumptions, recounting success stories of adult patients who visited him on an outpatient basis, having survived thirty-two aspirations for chronic and recurring abscesses over a two-year period.[85]

Joel E. Goldthwait, a graduate from Harvard Medical School of the class of 1890, aimed to broaden the scope of US orthopedic surgery, staking a claim in adult care by extending the same child-saving reform measures to working-class laborers. While interning at both Boston City and Carney Hospital under the tutelage of Bradford, Goldthwait encountered many working-class patients whose orthopedic conditions had the potential of permanently keeping them out of the workforce. Some patients came to him having sustained acute orthopedic injuries while on the job. Others came with chronic conditions—namely, adults who had survived childhood rickets or bone tuberculosis—looking for ways to get their bodies to conform to the new factory speed-ups, an industrial change that often proved to be too demanding on the chronically disabled.[86] One such case was a local thirty-year-old seamstress, who because of childhood rickets suffered from chronic slipping of both knee caps and thus found it difficult to keep pace with her co-workers. After a tendon release and rehabilitation, the patient was able to "work in a factory, able in every way to meet the demands made upon her."[87]

Goldthwait had a deep awareness of the changes occurring in the industrial workplace at the turn of the twentieth century and knew that these transformations had a direct bearing on his practice. In the new assembly line factories of the early twentieth century—where Taylorism reigned and speed-ups were the norm—laborers had to stand in one place all day while conveyer belts and machines moved around them. These changes in the actual physical conditions of the factory floor resulted in serious repercussions for the laborer's body, as static standing often led to more wear and tear on the joints than did moving and lifting. The new material conditions of Taylorism brought about mass strikes, as well as individual worker unrest, most frequently expressed in the doctor's office, where it was hoped that tired and injured bodies could be cured.

At a time when managerial capitalists were looking for ways to eliminate unwanted workers, namely, children, women, minorities, and the disabled, because of a bloated labor supply, Goldthwait insisted that it was better to get disabled men and women back to work than to allow them to become, in his words, "hopeless derelicts."[88] This goal of keeping disabled adults in the workplace instead of on the streets, begging for money, was most succinctly stated by Goldthwait's Boston colleague, Robert Osgood, who contended that orthopedic surgery could make all American working-class men into "happy, productive, wage earning citizens, instead of boastful, consuming, idle derelicts."[89]

Such statements went to the heart of Progressive Era concerns about the problems of poverty, the Civil War pension system, urban street beggars, and worker unrest. Unlike eugenicists who wanted to sterilize the unfit and legislate the removal of "defectives" from city streets, orthopedists did not shun the disabled. They could not, for a significant number of orthopedists were disabled themselves. Beyond the more immediate personal concerns, however, orthopedic surgeons possessed grander visions that their work could bring about social uplift. And so they aligned themselves with Progressive Era reformers who wanted to clean up the nation's cities, end political corruption, and create a safety net for industrial laborers.

Goldthwait's medical program, which he called "aftercare," won support among Massachusetts legislators, philanthropists, and fellow physicians. In 1899, the state of Massachusetts allotted $10,000 for the establishment of an orthopedic clinic devoted exclusively to crippled adults. This money went to the construction of a medical-mechanical department at the Massachusetts General Hospital, an outpatient clinic that opened in 1904 with Goldthwait as its first chief. Five years later, backed by the money that Goldthwait had raised from donations made by friends, patients, and colleagues, the hospital opened an inpatient ward for disabled adults, mirroring the layout, care, and vocational training that existed in the hospitals built for crippled children during the late nineteenth century.[90]

When war first broke out on the European continent in 1914, orthopedic surgery had not been a recognized military medical specialty in the United States, nor in most of the belligerent nations engaged in war. This fact would slowly change as Germany, Austria, France, and eventually Great Britain experienced such massive casualties that replenishing the front lines became nearly impossible. Introduced first in Germany and France in the wake of such disastrous battles as the Somme, in which armies endured upwards of 60,000 casualties a day, orthopedic hospital schools and vocational "curative" workshops became part and parcel of each nation's military defense strategy. These new programs in orthopedics promised not only to make moderately wounded men well enough to reenter combat, but also to train the disabled in weapons manufacturing, so that even the severely maimed could be made useful, working at jobs on munitions production lines. Other belligerent nations quickly joined in the rehabilitation project, reasoning that the war needed to be fought on both front lines and assembly lines. Ultimately this demand for functioning male bodies made rehabilitation seem like a necessity to the success of any modern state at war.[91]

Enthused by the gains that orthopedic surgery was making as a result of the Great War, Goldthwait believed by 1915 that the United States "would ultimately assume some part in the struggle," even though the Wilson administration maintained an unmistakable position of neutrality in the early years of the war.[92] Convinced that US involvement was imminent, Goldthwait created an "orthopedic war preparedness committee" in 1916, a group that would inform both the AOA and the American Medical Association of the growing need for military orthopedics.[93]

While at first glance Goldthwait's actions appear to be those of a warmonger, he was in all likelihood responding to the fact that many US medical personnel were anxious to assist in caring for the war wounded well before America officially entered the war. Along with doctors and nurses who volunteered with the Red Cross, certain American universities organized medical units to travel abroad. Most pertinent for Goldthwait was the surgical unit that Harvard University sent in 1915 to the American Military Hospital (originally established by American expatriates who were living in France prior to the war) in Neuilly. Harvey Cushing led the unit with Osgood as the representative for orthopedic surgery.[94] Subsequent Harvard medical units, as well as a Chicago unit, traveled to different areas of France to work at British hospitals, all at the Crown's expense. In addition to civilian involvement, the US Army's Medical Department sent its own convoys of doctors and nurses from the very outset of the war in order to gather information and assist in planning for the care of sick and wounded American soldiers, if the situation arose.[95] One outcome of these early envoys was the publication of the army's Medical War Manual Number 5, titled "Lessons from the Enemy: How Germany Cares for Her War Disabled," which emphasized the place of orthopedic hospital schools and workshops in the rehabilitation of the Kaiser's troops.[96]

Besides letters from Osgood, a crucial source of information about wartime orthopedics came from Sir Robert Jones. Jones and Goldthwait had become close colleagues and friends during the first decade of the twentieth century. In 1904, when he was president of the AOA, Goldthwait invited Jones to participate in the association's annual meeting, a request that Jones would not fulfill until 1907, when he attended the twenty-first annual meeting held in Washington, DC. Their relationship grew out of the fact that both men found themselves moving in parallel, if opposite, directions in their careers. Whereas Goldthwait, reared in pediatric orthopedics at the Boston Children's Hospital, was moving ever more in the direction of adult care, Jones began taking an increased interest in crippled children after years of treating injured adults. A nephew of famed Liverpool orthopedic surgeon Hugh Owen Thomas, Jones began his career at the age of sixteen, under the tutelage of his uncle, treating "shipwrights, ironworkers, boilermakers and dockgatemen."[97] At the age of thirty-one he began a casualty service, treating injured laborers who were building the Manchester-to-Liverpool Ship Canal during the years 1888 to 1893.[98]

Jones became interested in the orthopedic care of the crippled child largely as the result of Agnes Hunt, a British nurse who, as a result of her own childhood disability, established an open-air children's hospital in 1900 at Baschurch in Shropshire. After he performed several operations on Hunt herself, she asked Jones to became a consulting surgeon for the Baschurch home in 1904. Just as in the United States, British social welfare reformers of the late nineteenth and early twentieth centuries had taken up the cause of the disabled child, believing that the crippled would be best served in "labour colonies" where they could learn to become self-sufficient through an education in skilled labor. But whereas in the United States orthopedic surgeons spearheaded the movement of curing the disabled child through convalescence and vocational education, in Britain female social workers and nurses stood at the forefront.[99]

In the fall of 1914, the British War Office mandated that Hunt convert the Baschurch hospital for children to a center for wounded soldiers.[100] In early 1915 the War Office designated the Alder Hey Poor Law Infirmary outside of Liverpool as a hospital for reconstructive surgery under Jones's control.[101] True success, at least from Jones's point of view, did not come about until 1916, when the army medical services named him chief inspector of the Division of Military Orthopedics and granted his wishes for a centralized rehabilitation hospital and "curative workshop" at Shepherd Bush in London.[102]

Jones won the ear and purse of the British government and military by contending that through reconstructive orthopedic surgery, disabled soldiers would be shown to be "an essential part of the economic manpower of the nation, independent producers and wage-earners . . . not helpless dependents." Much like the Civil War pensions reformers in the United States, British social progressives worried about the economic and social effects of having more and more soldiers on the dole. With thousands upon thousands of disabled soldiers returning home, British War Cabinet Secretary Thomas Jones urged Prime Minister David Lloyd George in 1917 to provide continual support to Jones and Shepherd Bush lest the country be "saddled with thousands of untrained idle pensioners, who will ever be available as object lessons to which political wire-pullers can appeal."[103] Taking his cue from the Baschurch home, as well as from the multiple orthopedic aftercare hospitals in the United States, Jones, with the help of deposed King Manoel of Portugal, set up a system of rehabilitation that included medical-mechanical treatment, physical therapy, massage therapy, vocational training, and engineer workshops. According to historian Jeffrey S. Reznick, Shepherd Bush had as many as fifteen different engineering shops by the fall of 1917, each specializing in specific trade work, such as automobile repair, furniture making, tailoring, and cigarette production.[104]

What Jones lacked was medical manpower. British attempts to organize an orthopedic society during the late nineteenth and early twentieth centuries failed repeatedly, leaving English surgeons no other option but to join the AOA if they wished to find solidarity among like-minded professionals.[105] As soon as the United States declared war on Germany on April 6, 1917, Jones wired Goldthwait and other AOA members, urging them to come to Great Britain to assist him in the rehabilitation effort. US Surgeon General William Gorgas immediately endorsed Jones's request, telling Sir Alfred Keogh, the director general of the Royal Army Medical Corps, that he wanted to give "Dr. Goldthwait a general supervision over the orthopedic work . . . in this country in the same way that Robert Jones supervises this work, under your command, in England."[106] The "Goldthwait Unit," composed of Goldthwait and twenty other orthopedic surgeons, arrived in Liverpool on May 28, 1917, with the dual purpose of fact-gathering and providing aid. Less than a month after his arrival, Goldthwait already had impressive statistics to share with Gorgas. "Of the first 1350 cases seen at Shepherd's Bush," wrote Goldthwait in his report, "1000 were discharged back in the Army . . . and the remaining 350 were all discharged better prepared both as to the function of [their limbs] and [in their] industrial training."[107]

From a military standpoint, the benefit of orthopedic surgery was that it conserved men for the front lines. The conservation movement ran wide and deep in the United States during the early twentieth century, spanning the fields of hydrology, forestry, geology, anthropology, and industrial management. Come the First World War, it appealed to the military as well. Progressive Era conservationists envisaged, historian Samuel

P. Hays writes, "a political system guided by the ideal of efficiency."[108] Such a system, Hays maintains, was to be dominated by "technicians who could best determine how to achieve" such efficiency. In the case of human conservation during wartime, orthopedic surgeons became the technicians who would manage and engineer the soldier's body.

But orthopedic surgery as articulated by Goldthwait and Jones promised more than wartime conservation, for both men had their sights on an improved postwar society. Consider Goldthwait's initial recommendation to Gorgas, in which he outlined how US rehabilitation programs should take shape, based on Jones's example. "Hospitals at home should be fully equipped with curative workshops," he wrote, for with "such training many of the men [will be] able to earn *more* than before the War."[109] The audacity of this statement did not inhibit Goldthwait, for he indicated full awareness that policymakers in the United States were primed to dismantle the then-current veteran pension system: curative workshops "leave the man more independent rather than more helpless or dependent financially than before his injuries which is quite the reverse of the usual former [i.e., Civil War] condition of the wounded." In short, with this letter, Goldthwait articulated a vision of how orthopedic surgery could be used as a means to break from the past—a way to circumvent an injured soldier's postwar demand for a pension. In a 1941 retrospective view of his wartime work, Goldthwait explained that orthopedic surgery and curative workshops grew out of the deep, yet everpresent fear of postwar uprising and revolt. Knowing that many of the men drafted into war were the same men engaged in labor strikes, Goldthwait reasoned that "each of the individual wounded men represented a center of unrest." "And . . . unless something could be done to improve their condition," he concluded, "these individuals would become centers of revolution."[110]

Goldthwait returned to America in August 1917 with the army Surgeon General's Office having already begun the process of creating a Division of Orthopedic Surgery. The creation of a military division devoted to orthopedic surgery drew the support of the biggest names in the field. The orthopedic advisory committee to the surgeon general consisted entirely of former presidents of the AOA. Goldthwait was joined by fellow Harvard medical school graduates—all Bradford protégées—Elliot Gray Brackett, who chaired the committee, Dr. David A. Silver, and Dr. Lovett as member and liaison to the National Defense Council. Dr. Davis also participated on the committee and would have volunteered to join the military if he had not been too old to do so (he was sixty years old at the time). In addition to Davis and the Boston contingency, A. H. Freiberg, who received his degree from the Medical College of Ohio and practiced in Cincinnati, took charge of formulating standardized courses in military orthopedics for all nonspecialists in the Medical Corps.[111] While the advisory committee worked on home-front matters, orthopedic surgeons Osgood and Nathaniel Allison, both of whom had been abroad working with the Harvard Unit, were transferred from France to the British Medical War Office in London to serve as deputies to Jones and the British orthopedic service.[112]

From the outset, orthopedic surgeons intended to make themselves central to US military medical care, not specialists who would remain on the periphery. Surgeon General Gorgas had given them great leverage, mandating that the director of Military Orthopedics would be responsible for the care of orthopedic conditions, developing "all necessary organization[s], including special hospitals and their equipment . . . both at home and abroad." Brackett, who received the official designation as director of military orthopedics on the home front, took the definition one step further, contending that orthopedic care of the soldiers was to "begin at the time of injury and to be carried on continuously until the soldier is returned to active duty, or, disabled, is returned to industrial life."[113] The Division of General Surgery, the older and better-established military surgical division, felt threatened by orthopedic surgery's newfound power. Dr. Charles W. Mayo, a general surgeon who had always taken an interest in orthopedic matters, protested the creation of the Division of Orthopedic Surgery, as did famed Philadelphia surgeon William W. Keen. In response, Goldthwait and his colleagues insisted that, in following the example of Jones and the British, it was obvious that the two branches of surgery had distinct goals and practices. Goldthwait wrote to Mayo:

> The chief difference between general surgery and orthopedic surgery is that with the latter the operations that are frequently required are only incidental to the general treatment and can undoubtedly be performed by any reasonably clever surgeon. Following the operation, the great work of the orthopedic surgeon begins, in attempting by means of special exercise, special apparatus, as well as a great many other details, to restore the damaged part to as nearly perfect condition as it is possible to make. This, of course, means many times, months of the most patient handling with special equipment, for which the original surgical division is of course in no way provided.[114]

In other words, Goldthwait wanted general surgeons to stick to the knife while orthopedic surgeons managed all other aspects of patient care.[115]

The task of caring for disabled soldiers from the front line to the as-sembly line necessitated, as one army surgeon put it, "radical control." Well-versed in Fredrick Winslow Taylor's *Principles of Scientific Manage-ment*, Goldthwait became chief of military orthopedics for the American Expeditionary Forces and arrived in France in October 1917, whereupon he immediately set to work not only on the construction of 35,000 hos-pital beds but also on the creation of a more efficient and standardized system of triaging orthopedic patients. In order to make sure that injured men did not go home with joint contractures—"unnecessary deformi-ties which the orthopedic surgeon would be forced to correct later"—he established "orthopedic stations" at every step of the injured soldier's journey from the interior battlefields along the Western Front to the northern shores of France, where ocean liners would carry them back home. Each station would have a "splint team" consisting of one ortho-pedic surgeon and two orderlies who would apply Thomas leg splints or Jones arm splints (devices invented by Hugh Owen Thomas and Robert Jones, respectively) immediately upon injury and after surgery. These standardized splints were applied and reapplied at every hospital from front-line field hospitals and rear-line base hospitals to transportation hospitals on trains and evacuation hospitals. This assembly line of care, Goldthwait maintained, was necessary to successful aftercare at home. Output informed the preliminary steps of care. The point was to create a skilled laborer; to reach that goal, it was necessary to ensure that limb functioning was conserved and deformity prevented.

The home-front orthopedic surgeons were the ones ultimately left with the task of building an infrastructure that would make the end goal of orthopedic surgery a reality. With Brackett at the helm, Davis, Lovett, Freiberg, and F. H. Albee (New York) took charge of recruiting and edu-cating additional medical personnel, setting up emergency orthopedic coursework, training that lasted six weeks. In addition to the task of creating standardized artificial limbs for amputee soldiers, Dr. Silver au-thored the Medical Department's manual on orthopedic surgery except for three chapters that were taken directly from Jones's *Notes on Military Orthopedics*. The largest task, however, was the construction of military reconstruction hospitals and outfitting them with the proper equipment as well as with qualified personnel, including such medical assistants as occupational and physical therapists. With no end to the war in sight, the US Medical Department estimated that four reconstruction hospitals would be needed to accommodate hundreds of thousands of injured US soldiers coming home in need of rehabilitation. These hospitals would house curative orthopedic gyms—with hydrotherapy, electrotherapy, and exercise equipment—as well as workshops and simulated occupational workplaces.

*   *   *

While the creation of orthopedic aftercare programs for injured soldiers was a significant departure from the pre–World War I veterans' pension system, the practice of rehabilitation had many precursors both at home and abroad. The orthopedic surgeons in charge of creating military reconstruction programs had first-hand experience with establishing vocational hospital schools for crippled children in the late nineteenth and early twentieth centuries. Such institutions served as blueprints in the minds of the men involved with the Division of Orthopedic Surgery. The hospitals that would eventually be used for the rehabilitation of US soldiers disabled from the First World War mirrored the Widener, Massachusetts General's medical-mechanical department, as well as the curative workshops at Shepherd's Bush.

For the orthopedic surgeons who had been involved in treating disabled infants and toddlers, moving from the crippled child to the disabled young adult soldier required no great leap in imagination or practical skills. By July 1917, the Selective Service Act, the legislation that allowed for a universal draft, extended the draft-eligible age downward to include 18 year-olds, not much older than many of the inhabitants of crippled children's institutes.[116] Moreover, as historian Seth Koven points out, dismemberment in war was "instantly translated into a loss of masculinity and a loss of full citizenship."[117] In other words, disabled men and disabled children both fell into the category of dependents, whether on the state or on local voluntary organizations. What animated the practice of orthopedic surgery and rehabilitation was the creation of *future* citizens: taking a maimed body and repairing it in the hopes that someday it would be put to use in the industrial workplace.

THREE

A New Female Force

The World War I rehabilitation movement mainly revolved around men: male medical professionals treating injured male soldiers in order to get them back into a male-dominated workforce. Indeed, according to the wartime "Creed of the Disabled Man," which rehabilitators produced and publicized, the very point of physical reconstruction was for a disabled veteran to become "a MAN among MEN in spite of the physical handicap."[1] As the creed makes clear, rehabilitation advocates adhered to an assumption common to the early twentieth century that to be a man meant that one was economically viable, a self-reliant producer of capital, and the breadwinner of the family. To become a man in this context, then, meant that a soldier's disability had to be conquered, at least in the workplace.

Although it may have been a man's world, women still played a key role in defining the intent and practice of rehabilitating male soldiers. At the same time that the army Medical Department created the Division of Orthopedic Surgery in August 1917, Surgeon General Gorgas signed an executive order to hire "women war workers" to assist the Division of Special Hospitals in the rehabilitation project. While dietitians headed up the mess halls and occupational therapists cared for men with neuropsychiatric conditions, so-called "physiotherapy aides" worked with orthopedic surgeons in the rehabilitation of soldiers who suffered from fractures, shrapnel wounds, and limb amputations.

While the other allied health fields of occupational therapy and dietetics had been established as women's work before the war—both had histories closely connected with the Settlement House Movement and society work—physiotherapy was a new, undefined field in the United States.[2] As such, physical therapy became a decidedly post-Victorian women's work that adapted to the new gender constructs put forth by the WRIA and the rehabilitation program that followed. Whereas more traditional women workers cared for patients at the bedside, physiotherapists worked in gyms and performed manual therapy with the goal of reshaping weakened and disabled male bodies, making men stronger and fitter for the theatres of war and industrial work.[3]

Physical therapy necessitated physical domination; it required women to control and manipulate maimed male bodies. Such an interaction between the sexes would have been virtually unthinkable in the century leading up to the First World War. But with the WRIA came an implicit redefinition of the disabled veteran as a citizen-soldier who was in the process of *becoming* a man. Whereas the Civil War pension system permitted disabled veterans to believe that they had demonstrated their manly worth during war, the WRIA, and especially the institution of rehabilitation itself, emasculated injured soldiers, insisting that manliness could (and would) not be achieved until a disabled veteran reentered the workforce, laboring as an able-bodied person.

This chapter uses the story of the rise of wartime physical therapy to demonstrate how the rehabilitation project relied on women to serve both as war symbols and workers, despite the fact that the two roles often came into conflict. For rehabilitation to succeed, its advocates contended, disabled soldiers had to will themselves to take the harder but higher road of self-reliance, rejecting handouts from the state, private citizens, and charity organizations.[4] One key motivator, rehabilitators believed, was the disabled man's desire to marry and support a wife.[5] Here rehabilitators extolled the symbol of woman as a helpless yet domestically supportive partner. In contrast to this ideal of a damsel in waiting, rehabilitation advocates wanted the actual women who treated disabled soldiers in the clinic to exert a "kindly discipline," a combination of care and control. Douglas McMurtrie—one of the most vocal proponents of instituting programs of rehabilitation for US disabled soldiers—warned that women war workers should never engage in "sentimental pampering" of disabled soldiers, which he defined as "unwise sympathy proceeding from the heart but not from the head."[6] Scientific charity—upon which the new rehabilitation program was based—made little room for "emotion, sensitivity, and personal connection."[7]

In this complex web in which domestic womanhood was seen as both the enemy and the modus operandi of reconstruction, female physical therapists spun a professional identity that capitalized on the belief that injured soldiers needed stern women to make rehabilitation happen. Unlike educated women of the nineteenth and early twentieth century who accepted their lot as the weaker yet more nurturing sex, physiotherapists created an identity for themselves as strong women who possessed specialized medical knowledge: women, that is, with brains and brawn. Thus, physiotherapists became role models of how women should respond to the war wounded in the new, post-pension era of rehabilitation.

Making Medical Rubbers

At a wartime Orthopedic Advisory Council (OAC) meeting on August 13, 1917, orthopedic surgeons Elliot Brackett, Harold Corbusier, Robert Lovett, and David Silver all expressed concern about how the US Army was going to organize a "massage and physical training corps."[8] No one on the council disputed the fact that such a group was needed. The orthopedists of the OAC agreed with the Goldthwait unit report that the US Surgeon General's Office needed to follow Robert Jones's rehabilitation program, copying the format of reconstruction instituted in the Great Britain.[9] A necessary component of the Jones system was physical therapy, a panoply of aftercare treatments that ranged from gymnastics, massage, and hydrotherapy to sinusoidal baths and pulley apparatuses.[10] With over 65 percent of British soldiers suffering from casualties that involved impairment of locomotor function, orthopedic surgeons relied on physiotherapists serving in the British army to assist in rehabilitation work.[11] Nurses, enlisted men, and women from the British Society of Trained Masseuses all engaged in rehabilitative therapy.[12] In Canada, then a British colony, the Military Hospitals Commission initially wanted to train men exclusively for the job of physiotherapy, but with thousands upon thousands of injured soldiers returning home, the demand for rehabilitation necessitated that women enter the workforce. In 1918, the Military School of Orthopedic Surgery and Physiotherapy opened at the Hart House at the University of Toronto, where over 250 women were trained to become physical therapists.[13]

Orthopedists of the OAC worried that the US Surgeon General's Office, like the Military Hospitals Commission, would want the physiotherapy corps to be all men, too. Indeed, during the initial months of planning for such a corps, the Surgeon General's Office advised the OAC to enlist workers for physical therapy through the regular army via the Sanitary Corps. Although historically the Sanitary Commission was a department originally founded and shaped by civilian women during the Civil War, men with military rank dominated the Sanitary Corps of the First World War.[14] Brackett feared that the Sanitary Corps would exclude female workers from physiotherapy because of the army's discrimination against women, refusing them rank. Thus, in an August memo responding to the surgeon general's suggestion, Brackett insisted that the command for both the recruitment and organization of a massage and physical training corps be turned over to the Red Cross so that "women could be admitted."[15]

As latecomers to the war, having experienced virtually no causalities when the plans for reconstruction were first being drawn up, the United States could have made physiotherapy a man's job. Given the fact that the goal of rehabilitation was to find disabled men gainful employment, it is noteworthy that the United States did not take the path of the St. Dunstan's Hostel in London, where blinded soldiers were retrained to work as masseurs and employed by the Crown.[16] Despite such examples, the US Medical Department, and orthopedic surgeons more specifically, opposed the idea of training men as physiotherapists.[17] Orthopedists never made their rationale for hiring female physiotherapists explicit. They knew, of course, that the military needed men on the front lines and could not spare too many of them for hospital work. They also knew that women were cheap. World War I physiotherapists received $50.00 per month without military benefits, such as life insurance. Perhaps most important, orthopedic surgeons knew that female assistants would pose less of a threat to their male-dominated professional authority than male physiotherapist would. In short, as one orthopedist put it: "The work could be best performed by women."[18]

But not just any woman would do. The ideal woman, OAC members contended, would have a background in physical education, a profession that until that time had not been part of medical practice. Such a woman, they believed, would be more of a drill sergeant than a bedside nurturer— a medical assistant who could stretch and manipulate heavy limbs rather than lend a sympathetic ear. When, on another occasion, the Surgeon General's Office suggested that the OAC train a group of nurses to become physical therapy aides, the wartime orthopedists refused.[19] Despite the fact that large numbers of physiotherapists came from the same socioeconomic background as nurses (especially public health nurses), one physician maintained that a physical therapist should be "college bred," having a four-year university degree rather than only several years at a training school, as was common among nurses at the time.[20] Another orthopedist maintained that few nurses "made good as skilled operators."[21]

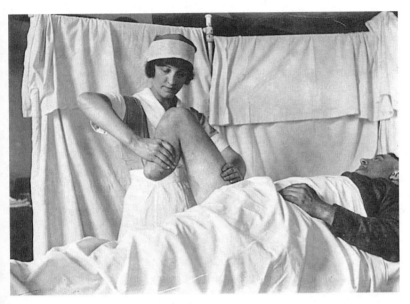

3   World War I physical therapist stretching an amputee soldier. Courtesy of the Emma Vogel
      collection, Otis Historical Archives, National Museum of Health and Medicine, Washington,
      DC.

The OAC's preference for physical educators over the more medically established profession of nursing arose from influences both outside and inside the practice of orthopedics. Rhetoric coming from outside medicine among lay rehabilitation advocates suggested that nursing could do more harm than good in the rehabilitation of disabled soldiers. Rehabilitators worried that a soldier, when showered by acts of womanly kindness and care, would lose the will to get well, failing to overcome his disabled state. "The man returning from the front deserves the whole-hearted gratitude and respect of the Nation," McMurtrie maintained, "but to spoil and pamper him is an ill-advised way of meeting the obligation."[22] Again, the fear in the minds of men such as McMurtrie was that the First World War would produce another generation of Civil War veterans, men looking for something (a national veterans' pension system) or someone (a concerned woman) to take care of them; the goal was to encourage them to take care of themselves. In order to convey his point that such womanly care, whether in the form of state welfare or actual personal acts, would lead to the decay of disabled veterans, he drew on analogies from childrearing, saying that while mothers wished the best for their children, they did "not manifest affection by spoiling the [child's] digestion with a . . . surfeit of candy." Likening the exhibition of womanly sympathy to that of giving candy to a baby, he concluded by asserting that responsible mothers do not give into their child's desires, but "rather . . . exert a firm but kindly discipline."[23]

Orthopedists similarly conflated nursing with overly sympathetic care and thus hired female physical educators to work as clinical assistants. In the decade leading up to the war, Goldthwait and his colleague Dr. Brackett offered summer courses on "corrective gymnastics" under the sponsorship of Harvard Medical School, specifically catering to women attending the Boston Normal School of Gymnastics. In these courses, female physical educators learned specialized exercises to treat medical conditions such as scoliosis, flat feet, and postural deviations. Goldthwait and Brackett, in turn, hired the graduates of these summer courses to work as so-called gymnastic assistants at the Massachusetts General medical-mechanical department as well as in their own private practices.[24] Lovett also hired a handful of female physical educators to assist him at the Boston Children's Hospital. During the 1916 polio epidemic, his gymnastic assistants performed muscle reduction, corrective exercise, and massage on the children who contracted infantile paralysis.[25] Women gymnastic assistants thought of their field as an "applied science." In their courses on anatomy, physiology, gymnastic exercise, and hygiene, students employed an arsenal of scientific instruments—including microscopes, skeletons, and anatomical charts—to better understand the effects of physical exercise.

When Goldthwait became chief of the Army Medical Department's Division of Orthopedic Surgery, he immediately appointed his long-time gymnastic assistant, Marguerite Sanderson, to the position of supervisor of reconstruction aides. Sanderson graduated from the Boston Normal School of Gymnastics with a degree in physical education in 1903. In 1913, she, along with two other physical educators, founded the Boston School of Physical Education and served on its board of trustees along with Goldthwait and Brackett. From then until the war, she worked for Goldthwait in his own private practice.[26] As chief supervisor of the reconstruction aides, Sanderson became the role model for how the first generation of physiotherapists should behave and what kind of education they should receive. The women who served under her found her to have a "gift for magnetic leadership." While they considered her to be a "disciplinarian," they also felt great affection for her, calling her "Sandy" instead of Ma'am or Miss Sanderson. One aide recounted that before her unit was shipped overseas, Sanderson always advised them to get "proper rest and wear warm underwear."[27]

When the Office of the Surgeon General finally agreed to adhere to the orthopedists' vision of what kind of women should become physiotherapy aides, a decision not made until the beginning of 1918, the Army's Medical Department looked to the nation's physical education programs for a supply of possible recruits and as sites where war emergency courses in physiotherapy could be conducted. By the spring of that year, six women's physical education schools offered their facilities to the Medical Department: in Oregon, Reed College; in Michigan, the Battle Creek Normal School of Physical Education; in Connecticut, the New Haven Normal School of Gymnastics; and, in Massachusetts, the American School of Physical Education, the Boston School of Physical Education, and the Prose Normal School of Gymnastics.[28] Wartime physiotherapy programs served as a crash-course initiation into military medicine. With physicians serving as medical directors, commanding the organization and content of the coursework, physiotherapy students learned anatomy in cadaver labs. Also part of the curriculum was "military orthopedics," where instructors would review basic wartime diagnostic groups, such as simple and compound fractures, ankylosed (i.e., adhered) joints, motor and nervous disorders, and amputations. And last, physicians taught physiotherapy recruits the theory and practice of applying deep tissue heating devices. Before graduating, physiotherapy aides had to demonstrate proficiency in electrotherapy and hydrotherapy, as well as the use of a wide array of assistive devices, including orthotics and prosthetics.[29]

The war emergency course instituted at Reed College became the largest program of its kind, producing the highest number of wartime physiotherapy reconstruction aides. By the summer of 1918, Reed accepted more than 200 women for their first four-month war emergency course. A typical day for a student included recitations of an "oath of duty," in which the recruit would pledge her allegiance to the United States, its armed services, and its soldiers. Aside from these character-building activities, students spent up to eight hours a day in classrooms and labs, while devoting two additional hours to calisthenics. "Every day," Reed student Marguerite Irvine recounts, "we had to run around the track after a solid hour of floor gymnastics and come back to the chinning bar for 10 pull-ups before we could don our bathing suits and dash for the icy cold lake."[30] Another Reed graduate claims that most of the 200 entrants in the first class "stuck it out to the end, although a number dropped out from sheer exhaustion."[31]

Calisthenics served the dual purpose of improving fitness and, more important, defeminizing physiotherapy recruits. Take, for example, the military squad song "Do or Die" that Reed students composed and sang in honor of one of their medical military commanders, Dr. Beach.

Yes, I'm ready to do or die, Dr. Beach, I'm ready to do or die,
You may make me chin beyond my reach or springboard ten feet high,
My chest is the frontest [sic] part of my frame, my tummy is flat and low,
And when I jump and kick myself, I always land on my toe.

I'm willing to do or die, Dr. Beach, I'm willing to do or die,
I'll cut up frogs and cadavers too, and the 8 hour law defy,
I'll dress my hair in minutes two, and dress in moments five,
With any old meals I'll gladly do, as sure as I'm alive.

I'm ready to do or die, Dr. Beach, I'm going to do not die,
I'll promise that I'll not flirt at all, or even wink my eye,
My relationship shall be business-like, the officers I'll not see,
And I'll look down so modest like, when a sergeant looks at me.[32]

As the lyrics make clear, army officials wanted physiotherapists to mute their femininity—no fussy hair, clothing, or food. The ideal body type of a physiotherapy recruit was one of plumb-line verticality. She was expected to control her body language; her gestures were to be muted, her gaze turned ever downward.

While some rehabilitation advocates worried that women war workers might be too sympathetic for rehabilitation to take effect—a maternal figure who would coddle injured men like infants—others fretted over the possibility of an overly sexualized woman who would become too intimate with her male patients.[33] This fear was particularly acute when it came to physiotherapy aides because in addition to taking coursework in anatomy and heating agents, physiotherapists learned techniques in massage.

Although massage had been part and parcel of the healing arts for centuries, it sparked great controversy among middle-class profession-als during the late nineteenth and early twentieth centuries. Massage underwent a revival of sorts during this period, with Swedish protégées of Per Henrik Ling (1776–1839) immigrating to Great Britain and the United States in order to teach the Anglo-Saxon world the methods of the so-called Swedish Movement Treatment.[34] Feeling inundated by the sharp uptick of massage therapists, the *British Medical Journal* (*BMJ*) de-nounced in 1894 the rise of unregulated "massage shops" in London, places that were usually found "in [the] clubland or immediate vicinity," where unscrupulous women treated male "patients who were not really ill."[35] The "Scandals of Massage," as it become known in the pages of the *BMJ*, echoed late Victorian concerns that massage parlors served as fronts for brothels.

In order to quell the fears voiced about distasteful massage practice, a group of English middle-class women formed the British Society of Trained Masseuses (BSTM) and made certification, education, and registration of all massage therapists a requirement for practice. To completely decouple the linkage between prostitution and massage, the BSTM refused male membership as well as male patronage. The BSTM prohibited general massage for men except in "urgent cases" or at a "doctor's special request."[36] Accordingly, the clientele for massage therapy moved from "young men about town . . . [making] a tour of the establishments," to middle- and upper-class women who could afford a private therapist for many hours of the day.[37] The BSTM, in other words, raised the moral status of massage, making it a kind of body work that respectable middle-class women could perform without the overtones of impropriety.[38]

Americans never became embroiled in a massage scandal, but the practice still raised the same kind of social and ethical concerns as the *BMJ* expressed, especially once the United States entered the Great War. Rumors abounded in the United States about how certain European nations, such as Austria, were losing the war because large numbers of soldiers were dying from venereal disease. To keep US men away from brothels and prostitutes, the War Department created a Commission on Training Camps Activities (CTCA) and mounted a widespread antivenereal disease campaign, a large component of which was to socialize soldiers to associate masculinity with sexual purity and abstinence.[39] The CTCA set up YMCA cantonments and bible classes as a way to provide more wholesome entertainment. Using these venues, the CTCA held frequent lectures that aimed to dispel the traditional assumption that men had natural, unbridled sexual passions that needed to be met in order to maintain their health and masculinity. The CTCA, historian Allan M. Brandt writes, "sought to invent for the troops a concept of virginity and chastity equivalent to that demanded of women."[40]

In a real sense, the US Army's recruitment of female physical educators to massage injured male soldiers during the First World War ran contrary to the agenda of the CTCA and, on a more general level, the accepted norms, practices, and beliefs concerning womanly conduct of the late Victorian era. While most physiotherapy recruits would have had some exposure to massage—most physical education programs included a brief introduction to massage—very few would have had experience in massaging men.[41] Ever since physical education programs arose during the mid nineteenth century, sex segregation permeated the methods, aims, and expectations of instruction.[42] Physical education schools for boys, for instance, arose during the Civil War in order to enhance male vitality and produce stronger warriors through gymnastics and drills. Girls' physical education, on the other hand, came about in response to post bellum medical concerns that highly educated women suffered from unnaturally high levels of "nervous tension"—a condition that not only resulted in ill health but also impaired a woman's ability to bear children. Thus, whereas men's physical education grew out of a desire to enhance and harness physical strength, women's fitness programs stemmed from cultural concerns that the weaker sex would become even weaker. As a system built on unwavering sex segregation, physical education programs only hired instructors who were the same sex as their students. Male physical educators instructed men to become more vigorous fighters and female instructors worked to create fitter mothers.[43]

In order to neutralize the perceived dangers that could result from training a cadre of young women to massage an army of young men, the Medical Department insisted that physiotherapists be known as "medical rubbers" rather than mere rubbers (a common early twentieth-century term indicating one who has some training in massage techniques).[44] By bringing rubbing under the umbrella of medicine and the alleged objectivity that accompanied it, orthopedic surgeons and the military thought they could successfully diffuse the potentially sexually charged interactions between soldiers and the women treating them, despite the unavoidable physical intimacy that massage techniques required. Female therapists were able to knead, mold, and stroke the male body without any suggestion of impropriety. Military courses in medical massage aimed to teach therapists how to see past the patient "as a sensual, aesthetic being," and instead see him as an objectified body, "a collection of mechanically oriented units."[45]

Putting Rubbers to the Test

No one knew what to expect from the women who graduated from the physiotherapy war emergency courses once they entered the field. Donner Jameson, a member of the first group of physiotherapy (PT) aides to be sent overseas to France in July 1918, recalled that many "medical men didn't know what we were or how to use us."[46] Another physiotherapist stationed at Savenay reported that she and her colleagues "acted as aides to the overworked nurses. We heated water for bathing . . . turned and re-turned soiled pillow slips, washed towels and carried drinks." Other accounts tell of "unfriendly and suspicious nurses," "severe and stubborn opposition from the regular army" medics, and disgruntled male patients who wanted nothing to do with reconstruction work.[47] With over 90 percent of the recruits having backgrounds in the nonmedical field of physical education, physical therapists seemed like imposters at the bedside.

Feeling unwanted, physiotherapy aides adopted a combative demeanor, especially with their fellow female war workers. Lena Hitchcock, an occupational therapist (OT) stationed in Chateauroux with a number of physiotherapy aides, described the PTs in the following way: "The Physios: Lavers [scolders], tall, middle-aged, disagreeable, always out of step." Hitchcock, who came from an upper-middle-class background found physiotherapists to be unrefined. In her unpublished memoir, she refers to one physiotherapist as "an elephant," another as "irritating and selfish," and yet another as the pinnacle of "sour, dirty womanhood."[48]

Although an outsider would not have been able to discern easily the difference between an OT aide and a PT aide—they shared everything from the same uniform to common living quarters—the women themselves felt a deep divide. Unlike physiotherapy, occupational therapy enjoyed the backing of women's charity networks and the settlement house movement. The practice began during the late 1880s when the Labor Museum at Hull House sponsored the Chicago Arts and Crafts Society, a group of Chicago reformers who engaged working-class men and women in woodworking and basket weaving to relieve nervous anxieties brought about by industrialization.[49] Out of this movement came Eleanor Clarke Slagle, known as the "Jane Addams of occupational therapy," who founded the profession.[50] During the war, occupational therapists brought to the battle lines this approach to easing physical pain, engaging patients in bedside activities to strengthen injured limbs and calm the minds of recuperating soldiers.

Wartime physiotherapists belittled the work of occupational therapy, characterizing it as a "pleasant handicraft that can be picked up in a few spare hours."[51] The worst insult, however, came from those physiotherapists who, despite their own pedigree, accused occupational therapists of being "society women," a slur created and adopted by the male framers of the reconstruction movement. Hitchcock remembered such an encounter with the physiotherapist Susan Hill, an assistant of Goldthwait's who was also stationed in Chateauroux. Before setting sail in July 1918, Hill, in a fit of fury, announced to the other physiotherapy aides that "Hitchcock possess[ed] almost every known undesirable trait . . . ; [that she was] the most contemptible of creatures . . . a society girl." Turning to Hitchcock, Hill continued, "you are spoiled, and the fuss your family made over you [before embarking overseas] was disgusting."[52] The commanding orthopedic surgeons encouraged this behavior among physiotherapists. "Too many [occupational therapy] departments," wrote one physician, "are under the auspices and control of well meaning ladies' boards, conducting the work as a social affair and quite independent of professional supervision."[53] In an attempt to rid so-called society women from reconstruction work completely, Goldthwait, while overseas, wrote a letter to the Surgeon General's Office claiming that although "the occupational aides have demonstrated their value quite as much as the so-called physiotherapeutic aides, all of them are really . . . physiotherapeutic aides and for that reason are being so considered."[54]

While physiotherapists benefited from the backing of their commanding orthopedic medical officers, they still had to gain the respect and trust of their male patients. Having had only three months of crash-course training, many therapists stationed both at home and abroad felt unprepared for what the work entailed. Ruby Decker, a graduate of the Normal School in Battle Creek describes treating her first patient at Camp Sherman, Ohio: "A parachutist on his first jump could not have been more apprehensive than I." Everything about her first treatment went according to protocol. She received the physician's diagnosis: ankylosis of the right wrist. Below the diagnosis, the physician wrote an order for treatment: hot bath, massage, exercise, and manipulation. But Decker lacked a basic knowledge of medical terminology and failed to understand that the soldier suffered from severe scar adhesions. "I could see that he had a stiff wrist," she writes, but "wondered how the 'ankle' got into the act." Her knowledge of the prescribed modalities was equally limited. "I'll never forget my first electrical stimulation," Decker recalls. "The instructions were printed inside the lid. I applied the indifferent electrode, turned on the switch, and then applied the active electrode on the designated motor point. Nothing happened! My patient had begun looking at the instructions, and started moving the metal bar back and forth. Success!"[55]

Because the technologies of electrotherapy and diathermy proved to be so complicated—and largely unavailable overseas—most therapists relied on massage as the primary form of treatment. According to postwar statistics compiled by the Physiotherapy Department at Fort McPherson, Georgia, as many as 7,297 massages were given in one month, compared with 2,164 treatments in electrotherapy. Every patient group, it was believed, could benefit from massage. In the case of amputation, all stumps

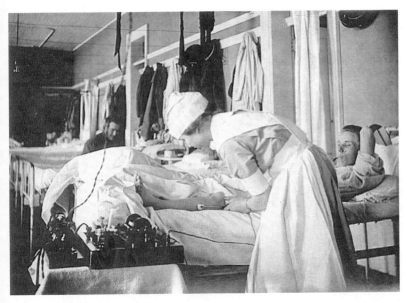

4   World War I physical therapist administering electrical stimulation to soldier-patient. Courtesy
    of the Emma Vogel collection, Otis Historical Archives, National Museum of Health and Medi-
    cine, Washington, DC.

received "massage . . . and compression bandages . . . to obtain the proper
condition of the soft parts." Therapists needed to keep stumps supple and
free from infection so that amputees could eventually wear prosthesis. Patients with osteomyelitis (infection of the bone, a potentially fatal condition before antibiotics) also received daily massages in order to "improve
local nutrition," "stir up any latent infection," and flush out the germs
through improved circulation to the affected area. Physiotherapists also
utilized massage for those men who suffered from peripheral nerve lesions, a condition that resulted in muscle atrophy and contractures of the
affected limb or face. In these cases, massage was used to maintain and
improve muscle tone until the nerve injury healed.

    Aside from the multitude of medical rationales, physiotherapists per-
formed massage because they believed that in comparison to other modalities of treatment it posed the least risk to the patient. More often
than not, injured soldiers coming home from the front suffered multiple causalities. To give but one example, Private S.H.H., wounded in
October 1918 by a high explosive shell, arrived in the United States in
December that year, having suffered from influenza, an amputated left
leg with an unhealed stump, osteomyelitis of the left shoulder with

"discharging sinuses," three large bed sores, and muscle wasting and contractures of his right knee because of poor bed positioning.[56] Cases like that of Private S.H.H. presented, one therapist wrote, "a puzzle to the medical and surgical staffs." In situations like these, the medical officers often ordered daily massage, believing that, at the very least, the treatment "could do no harm," and might even assist in recovery.[57]

Despite the demand for massage, most hospitals initially did not have facilities for its delivery. From the time of their arrival in September until December 1918, the Fort McPherson physical therapists resigned themselves to massaging patients at the bedside. Not until February 1919, three months after the conclusion of the war, when as many as 550 patients populated the physiotherapy wards, did the hospital build its first "surgical massage room," a 20-by-30-foot space, containing ten white pine massage tables, each 30 inches off the floor.[58] Once they got this new space, the Fort McPherson physical therapists expressed relief, pointing out that the room "aided effectively in establishing a professional atmosphere."[59]

The more physiotherapists relied on massage in their day-to-day practice, the more they argued for its further medicalization in an attempt to legitimize and objectify the act of middle-class women seeing, touching, and becoming intimate with the bodies of soldiers. In the spring of 1918, Sanderson delivered a paper at the annual American Orthopedic Association (AOA) tellingly titled "The Massage Problem." Sanderson thought that nurses, male patients, and nonorthopedic medical doctors might construe her work and that of her fellow physiotherapists as an illicit practice. Sanderson had reason to be concerned. The first leaders of the American Nursing Association (ANA), for instance, produced official statements throughout the early twentieth century distancing themselves from masseuses. Although day-to-day nursing care often included massage as a form of bedside treatment, ANA leaders argued that their young profession could not afford to be allied with a practice that seemed medically dubious (one ANA editorial likened massage to osteopathy, insinuating that it was nothing more than quackery) and, at the same time, suggestive of sexual misconduct.[60]

One way the physiotherapists legitimated their practice was to insist that "real" rubbing only occurred when the person receiving the massage was sick or injured. "The type of woman commonly known as a 'masoose,'" Sanderson told the AOA, "who, in civil life, devotes her time to massaging normal people employing massage as a luxury, is not the type of woman well qualified to follow the directions of a medical officer in the treatment of pathological cases."[61] It thus followed that the only women qualified to handle these so-called pathological cases were those who possessed "a thorough and practical knowledge of anatomy." Education separated the fraudulent from the authentic.[62] In this respect, physiotherapists engaged in a professionalization process commonly seen both inside and outside of medicine.

Another way in which physiotherapists made rubbing respectable was to mandate that the therapeutic encounter involve an element of pain. As Sanderson put it, the "patient's cooperation depend[ed] on the . . . *force* of suggestion given him by the masseuse." Feel-good, "luxury" massages, Sanderson maintained, did not have a curative effect. Massages that inflicted pain, however, had therapeutic purpose. Inflicting pain rather than pleasure protected physiotherapists against both the charge of impropriety and the rehabilitator's accusation that women were, by nature, sympathetic and thus ineffectual. By eliciting pain from disabled soldiers, then, physiotherapists complied with the greater vision of the rehabilitation project put forth by McMurtrie—a vision fueled by the fear that overly sympathetic women would ruin a man's prospect of successful rehabilitation.

A final way in which wartime physiotherapists maintained an air of professional legitimacy was by assuming command of drill and sporting events when not at the massage table. At Walter Reed Hospital, physiotherapists taught one-armed amputees to box, wrestle, and play baseball, just as they had been instructed in basic training camps.[63] Unlike regular boot camp, though, where able-bodied men competed against each other, disabled men in rehabilitation camps competed against women rehabilitation workers, most often the physical therapists.

In an era when physical education was strictly segregated along gender lines, such co-ed competitions would have been seen as highly irregular. Historian Martha Verbrugge points out that while other areas of American life and education had become more integrated during the early twentieth century, "physical education remained, only until recently, bifurcated by sex."[64] Many female educators during this period opposed having women engage in competitive sports, citing reasons ranging from social image to biological determinism. A woman's anatomical structure and physiological function, many argued, made her less able to perform intense forms of muscular activity and strain. Using such logic, female educators freely admitted that girls were "poorer throwers and catchers [in baseball] than boys."[65] In order to protect the "health and femininity" of women who played sports, educators of the 1920s designed separate rule guides (like half-court basketball) for girls who wanted to play "boy" sports, such as basketball and baseball.

No rule books existed, however, for co-ed events since they were considered culturally taboo. But neither did rule books exist for disabled men who wanted to engage in athletics. Thus when reconstruction workers began using sport as a means of rehabilitating disabled men, they had to adapt the rules and set up gaming competition to suit rehabilitation purposes. For instance, athletic directors had to decide whether or not disabled veterans could use assistive devices such as wheelchairs, crutches, and prosthesis when involved in track-and-field games. The demands of the particular sporting event combined with the physical condition (and mental willingness) of the disabled veteran often dictated whether or not assistive devices were allowed.[66] "Foot" races between one-legged men, for instance, were run with and without crutches depending on the men's ability to balance and hop on one foot. Events such as jumping hurdles (about three feet high) for one-legged men, however, almost always allowed for the use of crutches. For those who had lost both legs, athletic directors set up "chariot" competitions by which men would race against each other using their wheelchairs.

To a certain degree, having disabled men play against women in competitive team sports made sense, for both groups were marginalized, "handicapped" in their own ways. Both, for instance, required modifications to the "normal" versions of games that able-bodied boys played. While one-armed and one-legged men were hindered by war-maimed bodies, women, it was believed, had their own biological "handicaps." Menstruation, it was argued, lowered a woman's "vigor and performance" and was the chief reason why she could not compete against men.[67] Thus, setting up games between limbless men and able-bodied, yet inherently fragile, women leveled the playing field between the sexes.

At least in theory. Evidence from Walter Reed Hospital films and newspapers suggest that physiotherapists used such sporting activities to demonstrate their dominance over the disabled men they treated. One World War I military instruction video titled *Heroes All* makes it clear that women rehabilitation workers had a distinct advantage over their male patients when playing competitive sports. In the middle of the film, the viewer is introduced to the "PT Reconstruction Aides Team," a line-up of women ready to play the all-male "Amputee Team" in a game of baseball. The physiotherapists mercilessly defeat the disabled soldiers. Wearing light tops and knee-length bloomers, the PTs hit the ball effortlessly and advance swiftly around the bases to home, while limbless basemen and outfielders, in full dress uniform with empty arm and pant-legs pinned up, futilely hobble around on one foot, scrambling to get the ball. In contrast to the nimble athleticism demonstrated by the PTs, the legless men with no assistive devices or prosthetics struggle to keep up with their female competitors. When one legless man finally gets a hit, he haphazardly hop-runs to first, where he is tagged before he can reach the base. "Out on First!" reads the screen enthusiastically, and the player is escorted off the diamond by an umpire who dusts him off and gives him crutches so that he can make it back to the dugout without falling.

To be sure, the physiotherapists who ruthlessly beat the Walter Reed "amputee team" believed that they did so with the patient's best interest in mind. Rehabilitators and rehabilitation psychologists maintained that sport was necessary in order to heighten the disabled man's competitive zeal and ultimately affect his motivation to get well.[68] Walter Reed psychologist Dr. Bird T. Baldwin put it the following way: "Games had a particular psychological value in that they aided materially in the building up of a mental background of self-confidence, self-respect . . . and social cooperation with others."[69] But such games also worked in favor of the physiotherapists who, insecure about their place in both the medical delivery system and in the massage room, could demonstrate athletic prowess, take command of the baseball diamond, and essentially shame their disabled challengers into complying with the rehabilitation regime.

Among Walter Reed amputees, physiotherapists had a reputation for being "all chesty," often challenging (and beating) the nurses at a game of basketball.[70] One physiotherapist who worked during the 1920s recounted how some of her male patients would comment on her ability to perform such "hefty" work, wondering if all the manual stretching and intense sport might be "too much of a strain."[71] Pictorially, they portrayed themselves as women of rather large stature, arduously stretching heads and limbs of male patients and providing manual resistance during exercise.[72]

Because their treatments frequently elicited pain and shame in their patients, physiotherapists stood apart from other female war workers. Nurses, dietitians, and occupational therapists took on the more customary duties of female bedside nurturing; they fed, bathed, and cheered recuperating soldiers. Caroline B. King, a dietitian who served in the American Expeditionary Forces during World War I, expressed the nurturer point of view most succinctly when she spoke of the rewards of her work: "[I remembered] the bright faces of a whole ward full of desperately wounded boys . . . when I managed to give them something extra good, like lemon pies; but best of all my rewards was the name the boys bestowed on me—'Mother.'"[73] The physiotherapists, by contrast, actively avoided rhetoric that even hinted of maternalism.[74]

And yet, physiotherapists were not men, and they did not want to be. At every turn, they chose a middle road, characterizing themselves in both traditionally masculine and traditionally feminine terms. In army recruitment literature, they described their comportment as cheerful yet forceful and their touch as gentle but firm.[75] They believed, as one rehabilitation advocate put it, that the recovery of disabled soldiers was firmly "in the control of women." Women, and only women, could simultaneously adopt and embody both roles of disciplinarian and domestic ideal. It was thus the "responsibility" of women to "make [a soldier] believe he ha[d] a future."[76] The proverbial carrot held out in front of disabled soldiers was the image of a tender, caring woman waiting at home for their return.[77] Holding the carrot, though, were physical therapists who, through their tough demeanor, made sure that soldiers did not become too comfortable and reliant on bedside care in the rehabilitation process.

FOUR

Maximalist Medicine at Walter Reed

Replacing Civil War pensions with a robust system of re-habilitation necessitated not only a new cadre of medical and paramedical personnel but also the construction of uniquely designed hospital spaces and equipment. By May 1917, the Medical Department had already secured fund-ing from the War Department for the construction of new hospitals, a need made immediately apparent by the fact that upon the declaration of war, there were only 9,530 beds for an army that was predicted to swell 40-fold, numbering 4 million men in total. By the war's end, the army had spent approximately $2.5 million building 149 hospitals with over 100,000 beds.[1]

A large portion of the money went to the expansion of home-front general hospitals, sites where the new military programs in rehabilitation would take place. In a letter to the Secretary of War, Surgeon General Gorgas made clear the Medical Department's commitment to the new rehabili-tation ideal, saying that a man under his care would not be discharged until "he [had] attained [a] maximum cure and . . . is able to follow a useful occupation."[2] This "maximalist ap-proach to care," as historian Sanders Marble has phrased it, had repercussions on the entire military medical delivery system, from the hiring of civilian educators, machinists, and female paramedical personnel to drastic adaptations concerning the general hospital's material existence.[3] Tradi-tionally, military medicine focused primarily on acute care (largely wound management) and disease prevention—the basic premise was to keep soldiers alive. The turn toward rehabilitation, though, demanded more from the system and the soldiers, for returning the injured back to health—or even back to duty—was not enough. With the First World War, the military committed itself to the recuperation of "chronic invalids," those who, if discharged before fully rehabilitated, could potentially become a "burden on the State," pensioners like their Civil War brethren.[4] Building such a program required time and money: time not only for laborers to lay the bricks and mortar but also for long hospital stays. Instead of an immediate discharge with disability, the injured soldier of the First World War was expected to remain in the hospital for an average of six months, while his healthy comrades were being discharged and sent home to reunite with their families.

Surgeon General Gorgas designated two general hospitals to become permanent installations for rehabilitative care: Letterman General Hospital in San Francisco and Walter Reed General Hospital in Washington. Later in the war, the list of military rehabilitation hospitals would grow to 14, but Letterman and Walter Reed remained the flagship facilities during and after the war.[5] One reason that Letterman and Walter Reed rose to such prominence was that both had histories dating back to the Spanish-American War and were thus seen as permanent facilities—and indeed, as monuments-in-the-making—of the US military medical establishment. Aside from historical prominence, both facilities attracted the necessary funding and high-ranking medical personnel to keep the rehabilitation of maimed soldiers a centerpiece of the Medical Department's wartime agenda.

This chapter differs from most other histories of the hospital in the early twentieth century in its insistence that rehabilitation played just as much a role in the shaping of the modern American hospital as other, more commonly cited factors, such as the X-ray machine, the laboratory, antisepsis, professional nursing, and operative surgery.[6] While the "new" modern hospital of the early twentieth century often evokes images of marble-walled operating rooms, crisp white sheets, sterile shiny floors, and beds arranged in an orderly fashion to keep wards free of clutter, mess, and germs, there is another, less tidy side to the story. Rehabilitative care required that modern hospitals come equipped with exercise equipment, machine shops, orthopedic appliance shops, and hydrotherapy rooms, places where loud noises, wood shavings, debris, and puddles of standing water became the norm. Although standardized rehabilitative care had a military birthplace, it quickly became part of expected civilian practice in nonmilitary general hospitals. In 1926, famed Boston hospital architect Edward Stevens wrote, "Today the careful student of hospital architecture will not dare to plan his buildings without providing facilities for . . . physiotherapy [and rehabilitation]."[7]

To be sure, the rehabilitation units of Letterman and Walter Reed, as well as their civilian counterparts, would not have been possible were it not for advances in industrial and building technology (e.g., electricity, indoor plumbing). But the social reform driving and informing change in veteran welfare policy had more of an impact on the nature and structure of wartime hospital practice than anything else. In a 1917 article, orthopedic surgeon Leo Mayer argued that unless the army devoted resources to rehabilitative care, the long-term economic losses in pensions would be devastating. "America's war program," wrote Mayer, "must include not merely provision for establishing an army, but also for the care of those who are about to be discharged from military service and to re-enter civilian life." "Neither private charity, nor haphazard local organization," Mayer concluded in classic Progressive Era rhetoric, "must be allowed to care for these men, but the state itself must assume responsibility for them."[8] In short, Mayer and other like-minded rehabilitation advocates supported state medicine rather than state pensions, maintaining that money spent on the former—an ostensibly short-term fix—would save the country from the latter, long-term monetary obligation. It was this kind of social and economic calculation that gave impetus to the first military rehabilitation hospitals.

Early Years

Writing from Letterman hospital during the summer of 1918, Major Edward Rich told Colonel Elliot Brackett of the Surgeon General's Office that "it is hard to conceive of a more excellent situation for a general, reconstructive . . . hospital."[9] Letterman had been receiving similar praise since its establishment in 1899, a time when it was more plainly known as the military hospital at the Presidio. (It would not be dedicated to the famed Civil War army surgeon, Jonathan Letterman, until 1911.)[10] In his 1899 account of the opening of the Presidio's new medical facility, Dr. W. F. Southard maintained that, because of the hospital's location next to the sea, it was by far "one of the most healthful military posts in the world."[11] Built on the most northwestern corner of the 2000 acre military post, amidst a forest of over 2 million trees, the hospital sat near the entrance of the Golden Gate, where steamers and vessels carried troops to and from Manila during the Philippine-American War.

The hospital at the Presidio was modern in virtually every respect. The military used a modified pavilion design, featuring large, well-ventilated, open wards that were believed to prevent the spread of infection. First established in French hospitals during the 1789 Revolution and then solidified by the experiences of Florence Nightingale, who nursed soldiers back to health during the Crimean War, the pavilion design was well established by the 1890s.[12] Although Nightingale based her rationale for pavilions on the miasmatic theory of disease (i.e., that dirt, soil, and foul-smelling emanations caused disease), the design still persisted well beyond the discovery of germs, for until germs could be combated through the use of antibiotics, the hospital environment remained a crucial line of defense against infectious disease. During the late nineteenth and early twentieth centuries, hospital architects and workers believed that fresh air and sunlight were nature's most potent disinfectants, cleansing both the animate and inanimate objects housed within.[13]

The end result, for the Presidio at least, was a two-story wooden main hospital building, flanked by ten one-story pavilion wards on both the east and west side of the main building, with all wards facing southward, so as to maximize sunlight. Each individual 150-by-30-foot ward had hardwood floors and walls with rounded corners, "making them easier to clean" and to keep germ free. Ventilation—another crucial element of hospital design and disease control at the time—was "obtained through openings over each window, controlled by moveable glass frames . . . hinged at the bottom." The hospital was heated by steam and lit by electricity.[14]

From his station at the Washington Barracks, Dr. William C. Borden looked with envy across the continent at the Presidio. A year before the Presidio hospital was complete, Borden appeared before Congress in order to secure money for the construction of a new permanent hospital in the District of Columbia. Permanence had not been a common characteristic of US military medical institutions during the nineteenth century. After witnessing conditions of overcrowding and poor ventilation in the Georgetown Union Hotel Hospital in 1861, the US Civil War Sanitary Commission recommended that sick and wounded soldiers be cared for in tents and wooden shanties, structures that offered ample natural ventilation and could be easily abandoned and destroyed if infectious disease became rampant.[15] When Borden assumed his post at the Washington Barracks over thirty-five years later, little had changed. "At the time," he writes, "there were over two hundred patients, some of whom occupied the hospital building and others were in tents on the ground near it." As winter approached that year, Borden insisted that the military provide "better accommodations." A month later, two temporary wooden build-ings had been erected. "The whole establishment" Borden lamented, had a "make-shift nature."[16] By the end of 1898, Borden secured $300,000 in federal funding for the construction of a permanent medical struc-ture that would eventually be named Walter Reed in honor of his good friend and military medical colleague who suffered a premature death in 1902 soon after he was credited with the discovery of the cause of yellow fever.[17]

Like the Presidio, the central building of the Walter Reed hospital was based on the pavilion system, with wings placed laterally, all facing south. Borden insisted on wooden floors, not only because they were easy to keep clean, but also because they were the "pleasantest surface upon which to walk." Because of a delay of several years in the actual erection of the building, Walter Reed boasted cutting-edge technologies that superseded the Presidio. Rather than an exterior of wood, Borden had Walter Reed built with red brick and white stone facings. In addition, he advocated the installation of a newly invented mechanical ventilation system, known as a "plenum-vacuum system," which filtered incoming air by having it carried over the coils of hot-water pipes before being dis-tributed to the rooms internal to the hospital. Such a system ensured that air in the offices would be "changed three times a day," and four times a day in the wards.[18]

For Walter Reed to be a truly state-of-the-art institution, Borden main-tained, it had to be able to attract the brightest and best minds to mili-tary medicine. Well before civilian university hospitals became regular sites of medical education in the United States (a trend that would not take hold until after the 1910 Flexner Report), Borden urged the Medi-cal Department to wed the Army Medical School with the hospital so that military medical students would get ample clinical instruction. Convinced that Walter Reed would be a preeminent place for military medical education and research, he also lobbied to have both the Army Medical Museum, then located on the mall, and the Surgeon General's Library on site as well.[19] In many respects, his proposal to unite a school, library, museum, and hospital into one large, medical complex seemed outlandish at the time, leading one Congressman in 1905 to deride it as "Borden's Dream."[20] Although intended as a term of derision at first, "Borden's Dream" eventually became a rallying cry, and then a reality (of sorts) after the Second World War.[21]

Walter Reed, according to Borden, would stand apart from the more typical post hospitals, for the general hospital would treat patients with unique or highly complicated illnesses and injuries, not the "ordinary run of cases." From the beginning, Borden assumed that Walter Reed would admit subacute or chronic cases, patients who could not be treated by general practitioners in more remote places, without the resources of a working army college or research museum. "From a medical standpoint," he argued, "this makes service at [Walter Reed] particularly interesting, as difficult problems in diagnosis, prognosis, and treatment are constantly arising." According to Borden's dream, in other words, Walter Reed was to become a permanent place for medical specialties, an institution that would "require special appliances and special skill" to treat "obscure and difficult cases."[22]

That Borden was able to secure funding for a permanent hospital in the capital when the army was at war in the Pacific is a testament to his visionary abilities as well as his power of persuasion. He rallied support for his cause by evoking the historical significance of the area, pointing out that "in all previous wars in which the United States has engaged, troops in considerable number [had] been assembled in Washington and its vicinity."[23] At the same time, he appealed to the time-honored military maxim, "in times of peace prepare for war"—an adage that held great weight in a nation trying its hand at becoming an imperial power.

Borden's most powerful argument for the creation of Walter Reed, however, was that it would "reduce the pension list."[24] In 1898 Borden made his case before the Fifty-Fifth U.S. Congress for the appropriation of funds to create a permanent general military hospital in the District of Columbia using the following argument:

> The professional work of a medical officer of the army has a definite economic value, a value which can be accurately measured in dollars and cents. This arises from the fact that soldiers incapacitated for service on account of diseases or injuries acquired in line of duty receive pensions. . . . Consequently, if a medical officer removes any evident disability from an enlisted man . . . and returns him to duty, he saves the Government the amount of the man's pension which he would have received in case of discharge for disability, and, in the case of an officer, saves the Government an annual outlay to the amount of the officer's retired pay.[25]

At the Presidio, Dr. Southard engaged in similar kinds of argumentation, claiming that it was unjust for the federal government to "pay hundreds of millions for pensions," but spend so little on the Medical Department. "If there is one department more than any other which suffers from the paring process," Southard wrote, "it is the medical."[26]

At their very inception, then, both Walter Reed and Letterman General Hospitals were bound up with the Civil War pension debates that erupted on the political scene at the turn of the twentieth century. To men such as Borden, creating permanent military general hospitals was an institutional—and medical—solution to the problem of a bloated pension system. Better medicine, it was believed, would lead to better recoveries and fewer military discharges due to disabilities. And Borden had numbers to prove it. After only one year of service as commander of the hospital, he claimed that he kept "480 disabled men" from leaving the service. "At a pension rate . . . from $6.00 to $65.00 per month," Borden reasoned, he had saved the government that year "$133,065.48—the equivalent of three per cent interest on an investment of $4,435,516.00."[27] In essence, Borden wanted to show that within a year's time, Walter Reed Hospital had already paid for itself. Additional medical services would not cost the government more; rather, as Borden showed using the pension system numbers, it would cost the government less.

Great War Expansion and Diagnostic Essentialism

While Borden wanted to keep as many disabled men as possible off the pension list by returning his patients to military duty, the goal of the World War I rehabilitation project outstripped Borden's ambition in its belief that disabled men could be remade to fit back seamlessly into civilian life. Or, as the first wartime rehabilitation workers at Walter Reed put it, the purpose of reconstruction "was to help each patient . . . find himself and function again as a whole man—physically, socially, educationally, and economically."[28] Such lofty goals demanded that military general hospitals transform into institutions more akin to the Widener Memorial School for Crippled Children in Philadelphia and other such late nineteenth-century hospitals that specialized in the rehabilitation of crippled children. In other words, because of the antipension demand to get injured soldiers back into the civilian workforce, Walter Reed and Letterman became hospitals that not only healed but also educated its patients in on-site schoolrooms and industrial workshops.[29]

Rehabilitators maintained that getting an injured soldier from the front line to the assembly line required a multistep process that necessitated an interdisciplinary team of medical and nonmedical experts, as well as a variety of new hospital spaces and equipment. Broadly speaking, rehabilitation entailed three steps: first, soldiers injured during combat underwent repeated medical and surgical interventions, beginning with the stretcher-bearer who provided first aid on the front lines to the complicated system of base, field, and transport hospitals through which patients traveled before reaching a general hospital on US soil. Upon arrival at a general hospital, the soldier-patient who needed rehabilitation was enrolled in the second phase of rehabilitation—bedside physical and occupational therapy. Once a patient became ambulatory, he graduated to the third step of rehabilitation: vocational training.

Although rehabilitation proponents liked to emphasize the latter parts of the process, seeing vocational training as the linchpin to reconstruction, in reality most patients who ended up in the rehabilitation programs at Walter Reed and Letterman experienced months of overseas and domestic medical and surgical care before vocational therapy even began. Consider one wounded soldier, known in the records as "Lieutenant C." Assigned to the American Engineers Corps in France, he sustained a wound to his knee sometime in the spring of 1918 when a "shell fragment hit it and scooped a large hole in the knee-cap." Transported immediately to Base Hospital No. 9 in Rouen—controlled by the Cleveland's Lakeside medical unit and commanded by Dr. George Crile (later founder of the Cleveland Clinic)—he received a Carrel-Dakin treatment (a preantibiotic era method of cleansing wounds with chlorine), but to no avail. Within hours, Crile and his team amputated Lt. C.'s limb to stop the spread of infection. Weeks later, the same surgical team attempted to "fill in the great gap" that the shell had left in Lt. C.'s leg by grafting skin and tissue from another man's freshly amputated limb. After several unsuccessful attempts at using donor tissue, the team of surgeons excised tissue from Lt. C.'s "good leg" to make a graft for the injured leg. The tissue took, but Lt. C. writes, "it took six months for the area from which the graft was taken to heal."[30]

In between operations, Lt. C. spent time in a Balkan frame bed, a suspension device that orthopedists used to apply joint traction on patients with fractured and amputated limbs. Despite the fact that Lt. C. spent months in the Balkan bed, with one end of a pulley attached to his stump and the other end attached to bricks hanging over the end of the bed in order to keep the skin around his stump on a constant stretch, bone still protruded, making future artificial limb wear nearly impossible. In order to remedy the problem, Lt. C. underwent an operation to reamputate the limb at a hospital in Great Britain on his way to the United States. Finally arriving at Walter Reed Hospital in the winter of 1918, Lt. C.'s limb still remained unhealed. During that winter, he suffered from multiple bouts of internal hemorrhaging, leading his team of surgeons to perform two subsequent reamputations.[31] According to hospital statistics, Lt. C.'s experience was more common than not. By the end of 1918, the orthopedic surgeons at Walter Reed Hospital estimated that over 90 percent of amputees coming from overseas required further operative treatment—mostly reamputations—before the patients could be fitted with artificial limbs and begin rehabilitative care.[32]

Because of the number and chronicity of cases such as Lt. C.'s, the orthopedic wards in both Letterman and Walter Reed expanded rapidly during and after the war. Letterman Hospital saw its orthopedic ward capacity quadruple within one year, going from three wards in the spring of 1918 to twelve in the spring of 1919. Homeside general hospitals experienced the greatest demand for orthopedic services after the Armistice of November 11, 1918, a time when the army Medical Department began dissolving overseas medical hospitals, evacuating as many of the wounded back to the United States as possible.[33] By December 1918 at Walter Reed, for instance, 50 percent of all newly built wartime wards fell under the orthopedic service, totaling 39 buildings in sum, with each ward holding 25–35 beds.[34]

In both hospitals, orthopedists divided their patients into two main categories: patients with suppurating wounds resided in different wards from those with clean wounds (to avoid cross-contamination) and ambulatory patients lived separately from the bedridden. From there, orthopedists classified patients according to roughly six main diagnostic groups: amputees, bone injuries (fractures by explosive force or falls), joint lesions, nerve injuries (usually cases where nerves were completely severed because of explosive projectiles), tendon injuries, and last but not least so-called "static deformities," a category that usually included draftees who, upon initial military medical evaluation, were found to have flat feet.[35] Indeed, at the outset of the war—before home-front general hospitals received any of the wounded from war—Letterman designated an entire ward for "cases of weak, deformed or wounded feet needing shoe-corrections, plates, or other appliances."[36] For those draftees with flat feet, the process of rehabilitation began before they even saw combat.

The surgeon general's insistence that patients must be triaged according to physical diagnosis without regard to ethnicity, regionality, or race meant that, at both Walter Reed and Letterman, black and white soldiers were treated alongside on another. Although the army practiced Jim Crow, enforcing segregation between black and white troops, rehabilitation hospitals were racially mixed.[37] While at first glance the desegregation of white and black troops in an era of tense race relations might appear progressive, the Surgeon General's Office made it repeatedly known that it had "no intention . . . to settle the so-called Race Question."[38] The highest appointed black military official, Emmet J. Scott, the Special Assistant for Negro Affairs to the Secretary of War, campaigned throughout the war for the deployment of more black physicians in the army. According to Scott, there was "much dissatisfaction" voiced in the black press and within the NAACP about the "seeming disinclination on the part of the Surgeon General's Office to commission and utilize an adequate number of colored medical officers to minister the physical needs of the 400,000 Negroes who served in the Army."[39] Instead of being used for their professional expertise, most black physicians were drafted as privates. In addition to fighting against overt racial discrimination, black medical societies, according to Scott, "protested . . . the idea of permitting white physicians to serve in connection with colored units."[40] Black physicians wanted segregated medical facilities because they worried that black soldiers would not receive adequate care in the hands of white medical officers. According to one report from Camp Hill, Virginia, while white soldiers had ample hospital accommodations, sick black soldiers during a cold winter of 1917 "were huddled together . . . eighteen in one tent, without any wooden floors in the tents." Scott directly attributed the "abnormally high death rate" among black soldiers to the blatant negligence of the black man's health and basic needs by the Surgeon General's Office.[41]

When Scott sent a letter to the Surgeon General's Office inquiring about the medical rehabilitation of black soldiers, he received a predictable reply. C. R. Darnall Executive Officer of the Medical Corps wrote to Scott that disabled amputee soldiers would be sent to General Hospitals "without regard to color." "The policy of the Medical Department of the Army," continued Darnall, "is to treat colored soldiers in the same hospitals as white soldiers, and no separate hospitals will be established for the treatment of colored soldiers."[42] Again, while on the face of it this statement appears to be a moment of progressive thinking in terms of race, Scott and the wider black professional community considered it a defeat. So far as the records show, there were no black physicians in the Great War who were trained and practicing in orthopedic surgery and who would advocate for specialized attention to black soldiers. Instead, the Surgeon General's Office assumed that those black physicians permitted to serve in the medical corps should put their energies into combating contagious diseases—such as tuberculosis and syphilis—which were believed to be rampant among black troops and a direct threat to healthy white soldiers. In the white medical community, orthopedic injuries did not carry the same kind of racially charged cultural meanings or evoke the same degree of race-based fears and were therefore left to white surgeons to control and treat.[43]

5   Despite Jim Crow laws, Walter Reed General Hospital rehabilitated white and black soldiers in desegregated facilities during World War I; *Carry On: A Magazine on the Reconstruction of Disabled Soldiers and Sailors* 1 (May 1919): 5.

How Orthopedic Surgery and Rehabilitation
Changed the Modern Hospital

The institution of orthopedic surgery as a new military medical specialty during the First World War brought about a new material reality for the military general hospital. Not only did orthopedists demand roomier wards in order to accommodate Balkan frame beds and bedside wheelchair access, they also insisted on additional nonward space, places where they could engage in so-called "dry surgery," a practice that took place in the aptly named "appliance shop" where orthopedists, alongside hired mechanics, would repair shoes, cast limbs, and make braces.[44] At the outset of the war, neither Walter Reed nor Letterman had fully operational appliance shops. Evaluating Letterman's capacity as a rehabilitation hospital for the Surgeon General's Office in the summer of 1918, orthopedist Robert Osgood voiced disappointment with what was then passing as an appliance shop. Only the simplest of tools were found, "a forge and anvil" and the most basic "standardized [orthopedic] supplies and accessories." Given such paltry conditions, chief of Letterman orthopedics Dr. Leo Eloesser requested and received $1,200 in September 1918 to expand the shop so that it would include a lathe, an upright drill, a shear, a grinder, and a leather-sewing machine. By December of that year, Eloesser proudly reported that he had a fully equipped shop and had hired seven mechanics to work under him, three civilian "leg makers" and four military privates who specialized in metal and lathe work.[45]

In addition to the appliance shop, orthopedists at Walter Reed and Letterman oversaw the construction of physiotherapy buildings, spaces outfitted for hydrotherapy, mechanotherapy, electrotherapy, massage, and exercise.[46] At Walter Reed, the Engineering Corps subdivided the large 42-by-410-foot physiotherapy building into several 20-by-30-foot treatment rooms, designating each room by the therapeutic modality contained within. The massage room, for instance, sat adjacent to the "baking room" (devices used for electrotherapy were often referred to as electric bakers) and the electrotherapy room was situated next to the

6   An example of a fracture ward, with Balkan frame beds lined up on opposite walls, in an American military hospital in France. The United States Army Surgeon General's Office, *The Medical Department of the United States Army in the World War* 11, pt. 1 (Washington, DC: GPO, 1927), 624.

7   An example of a wartime orthopedic appliance shop, a rather messy and cluttered addition
to the new modern hospitals in both the military and civilian sectors. The United States Army
Surgeon General's Office, *The Medical Department of the United States Army in the World War* 5
(Washington, DC: GPO, 1923): 338.

hydrotherapy room. To supply the physiotherapy building and the orthopedic appliances shop with an adequate (and steady) supply of electricity, Walter Reed built and operated its own power plant, reportedly
the costliest item of all wartime expansion.[47] The largest physiotherapy
room—the mechanotherapy room—housed pulley weight apparatuses
and balance beams. Consolidating all the various forms of physical therapy treatment into one building, rehabilitators argued, made it possible
to "treat the maximum number of patients . . . by the minimum number
of personnel."[48] With all treatment technologies under one roof, a team
of 30 physiotherapists could treat as many as 550 ambulatory patients
per day.[49]

Surgical and medical work aside, the most crucial part of making Walter
Reed and Letterman rehabilitation hospitals was the creation of "curative
workshops." Early in the war, so-called curative workshops were little
more than maintenance sheds, places where carpenters and automobile
repairmen worked to keep the grounds, equipment, and infrastructure of
the hospital in good working condition. Walter Reed's first experiment
with curative work took place in the Lay Homestead (a site that served as a
lookout for Confederate soldiers during the Civil War), a small, two-room
house where the post's carpenter and family lived and worked.[50] Here, the
first group of disabled soldiers admitted to Walter Reed—most of whom

had sustained injuries on the home front as a result of accidents on trains or in camps—engaged in hammering, sawing, and varnishing as part of their medical treatment and recuperation. Likewise, Letterman's first foray into curative work took place in a "small shed near the hospital garage."[51]

This singular act of turning the working quarters of army post carpenters and mechanics into places of medical treatment demonstrates the degree to which rehabilitative medicine became insinuated into regular hospital practice. Hospital maintenance—and its employees—no longer served the mere purpose of keeping the physical infrastructure in good working condition. Rather, such sites would become places for medical cures, bringing normal, every day labor under the umbrella of medicine.

By the fall of 1918, the War Department allotted approximately $3,000 to each hospital for the purpose of constructing permanent curative workshops.[52] Walter Reed Hospital used the money to build six new wards, blandly designated by the numbers 93–99. Although from the outside the one-story wards were indistinguishable from each other—each measuring 24-by-144 feet and made of stucco hollow tile—they had a unique function keyed to the activities that took place inside. Buildings 93 and 94, for instance, housed the psychological laboratory—a place used more for measuring joint mobility than for counseling—as well as the "commercial department," where soldiers could learn touch typewriting, shorthand, and stenography.[53] Building 95, by contrast, became known as the "electrical department," where a mix of civilian and military instructors taught maimed soldiers Morse code and radio telegraphy. Here, too, soldiers could take courses in the engineering and drafting department learning techniques in mechanical architecture, topographical drafting, tracing, and blue-printing. While building 97 housed the "laboratory for artificial limbs," building 96 served as the curative machine shop and the "cinematograph department," where civilian instructors taught basic machine repair and classes on the latest in motion picture technology. Finally, building number 99, the "automobile school and shop," engaged soldiers in engine and tire repair.[54]

Although curative workshops provided an education that many soldiers lacked, the primary rationale for vocational instruction—at least as it occurred in the hospital—was medical. While the military hired civilian laborers and educators to perform most of the instruction that took place in the curative workshop, army medical doctors remained in charge of virtually every aspect of care. Military physicians, mostly orthopedists, prescribed curative workshop treatments much like pharmaceuticals. A wounded soldier with a contracture (limited motion because of scarring) of the right elbow, for instance, might receive a prescription to engage in sawing, an activity believed to enhance the flexion and extension of stiffened elbow joints. Similarly, physicians had soldiers with ankle injuries work on machines powered by foot treadles. The up-and-down pumping motion required by these machines was believed to be "the most rational and effective means of exercising joints and muscles" of the ankle.[55]

To understand better how vocational training became medicalized and a regular part of health care delivery during the war, consider the case of Private S.B. Sustaining multiple injuries in July 1918 at a battle in Chateau Thierry, the 26-year-old soldier arrived at Walter Reed in October that same year with "gas gangrene chest"—an infected, lacerating, granulating wound—as well as nerve injury and marked joint contractures of the right forearm. Once the chest wound had healed, S.B. began vocational therapy with his mornings spent in the curative "jewelry shop" and his afternoons farming. Although he had finished only seven grades of school and had worked as a farmer throughout his young adult life, both he and his treating physicians understood his work in the jewelry shop to be "more curative" in the sense that it not only improved his joint mobility but also gave him an education in skilled labor, providing him with a better chance of finding steady employment after his discharge. At the jewelry shop, S.B. busied himself with "metal work, specifically pulling copper wire in order to enhance [the] strength in [his] fingers." He started off slowly at first, only working fifteen minutes at a time in order to avoid fatigue. As his strength and stamina improved, S.B. began working with small filing tools in order to improve his ability to manipulate small objects. At first, S.B. used his left hand to assist him in handling these delicate tools, but by the time of his discharge he was able to master the task using only his injured right arm.[56]

Work therapy was by no means new to the medical profession. Before World War I orthopedists had been prescribing vocational training in hospitals devoted to crippled children since the late nineteenth century. At the same time, a handful of physicians in tuberculosis sanatoriums advocated work therapy, believing that a patient's immunological resistance was actually boosted through heavy manual labor.[57] British physician Dr. Marcus Paterson maintained that if tubercular patients worked hard enough, they would release their own tuberculin, a chemical byproduct that naturally improved resistance to the *tubercle bacillus* (a view which later became known as the autoinoculation theory).[58] In Paterson's sanatorium, patients would walk up to ten miles a day, carrying spades and shovels for digging.

An even earlier precursor to the curative workshop, however, existed in late eighteenth- and early nineteenth-century insane asylums. Working in Paris in 1778, Philippe Pinel set out to reform medical treatment for the mentally ill, overhauling a system based largely on corporal punishment and replacing it with asylums that would provide programs in "moral therapy." Through an orderly set of asylum activities, Pinel argued, those deemed to be mad—alcoholics, epileptics, argumentative wives—would "internalize the behavior and values of normal society" and eventually be cured.[59] The moral therapy approach was adopted in England by William Tuke (of the York Retreat) and finally in the United States by Thomas Story Kirkbride, superintendent of the Pennsylvania Hospital for the Insane in Philadelphia from 1841 to his death in 1883.[60] At the Pennsylvania Hospital, patients engaged in all forms of occupation, from Magic Lantern slide shows to farming and groundskeeping.[61] Although every asylum had a unique therapeutic approach, each abided by Pinel's principle that the purpose of moral therapy was "to break [the patient's] will . . . without causing wounds."[62]

Whereas Pinel's moral therapy existed in order to break a patient's will (i.e., making rowdy undesirables into well-mannered citizens), the rationale behind the World War I curative workshop was that it would help inspire the disabled soldier, sparking his will to work. In addition to the physical benefits that came from hands-on work, rehabilitators maintained, curative workshops provided a "psychological cure" for soldier-patients who were prone to "pension psychosis."[63] When first injured, Mayer wrote, "all of [the soldier's] thoughts revolve about the pensions, and he even resists the physician's efforts to cure him so that he may secure the maximum sum of money" from the government. Work, it was argued, would ignite a soldier's desire to get better, providing him with a cheerier outlook on life. Rehabilitators believed that disabled soldiers were particularly prone to mental depression, by which they meant a state of mind that manifested itself in self-pity, inactivity, and dependency on others.[64]

Once a man found work, it was argued, the rest of life—a wife, family, community involvement, and leisure—would naturally follow. Orthopedist Fred H. Albee, director of wartime orthopedics and rehabilitation at General Hospital No. 3 in Colonia, New Jersey, explained the importance of work as a life-source in the following way:

> Work, it has always seemed to me, is man's greatest blessing. Work and its attendant sense of achievement, of accomplishment is the most completely satisfactory

experience in human life. The day that is engulfed to the very edges in work is a good day. Perhaps it is the most enduring form of happiness. . . . Leisure is good, too, if it is a refreshment and a preparation for future labors. But leisure as a reward, an end in itself, a climax of living, a prize for a busy life—never![65]

Those in charge of instituting rehabilitation at Walter Reed and Letterman worried that without a curative workshop, soldiers would become too accustomed to a life of convalescence, and, by extension, a life of leisure. The Canadian Military Hospital Commission—from whom the US Medical Department consistently sought advice—put it more bluntly: "too much softness, in little convalescent homes, debilitated morale and added to the looming pension burden." W. M. Dobell, a member of the Canadian Hospital Commission likewise argued that there was "absolute unanimity of opinion [among the Allied Powers] that the influence of convalescent homes [was] bad." "Men are shown a different standard of living, from what they have been accustomed to," Dobell argued, "and one which they will probably not be able to maintain."[66]

This fear of convalescence pitted the new rehabilitation hospitals of Walter Reed and Letterman against the small already established system of National Homes for Disabled Veterans of Service, a federal network of institutions devoted to the long-term care of disabled Civil War veterans. Set up by the largely female-dominated United States Sanitary Commission (USSC), the post–Civil War National Homes reflected the practice and language of female benevolence, institutions that privileged domesticity, hominess, and comfort.[67] In 1865, when one USSC advisor, Frederick Knapp, suggested that soldier homes include workshops and schoolrooms, he received no support from his fellow USSC members. In the end, National Homes served as places of shelter for those disabled Union soldiers who were left homeless as a result of a death of a family member or the loss of a job. Most Union veterans who utilized the National Home system did so as "temporary or seasonal refuge." Most tenants, that is, "cycled between periods of institutional relief and periods of autonomy and employment."[68] And, most important, as the name implied, the institution functioned as a home, a place to rest, free from the demands and expectations that all adult, male citizens be gainfully employed, at all times of their life.

World War I rehabilitation, by contrast, became rooted in hospitals, not convalescent homes. At Walter Reed and Letterman, patients were expected to be active participants in the cure: actively engaging in their treatment, actively overcoming their disability, actively readying themselves to become self-reliant citizens.[69] In comparison with the National Homes, then, World War I rehabilitation was a decidedly aggressive approach to veteran welfare. A 1919 issue of the Letterman Hospital newspaper, *The Listening Post,* made clear its turn away from the supposedly feminine approach to the wounded soldier-patient. "Don't gush over him," the paper read, "or tell him what a hero he is nor weep upon his shoulder."[70] Most important, by instituting curative workshops, the hospital came to look more like a traditionally masculine place—an industrial machine shop—than the female realm of the home.

* * *

Although the USSC helped to establish National Homes for disabled Union soldiers, a few influential men in the organization's leadership opposed widespread institutionalization of disabled soldiers, fearing that the growing consolidation in state power would encourage veterans to become ever more dependent on the government. Speaking on behalf of the USSC shortly after the Civil War ended in the spring of 1865, director Henry Bellows argued that "the pension system [was] the proper substitute for military asylums and hospitals."[71] "The fewer the monuments to our martial law and our purely government regime," he wrote, "the better." Rather than "herding" disabled soldiers into institutions, Bellows maintained, the pension system "reinforced the role of family and community as the very center of American national life," making it possible for disabled soldiers to become autonomous, self-reliant individuals. Many politicians shared Bellows's viewpoint and complained about having to contend with the "disgruntled residents" of the already existent Washington Soldier's Asylum. "A day does not pass over my head," wrote one member of the 35th Congress, "when some one or more of these old men do not come hobbling to my room . . . making these complaints."[72]

Over a half-century later, Washington would be home to one of the largest military general hospitals for disabled soldiers and Progressive reformers would be taking the exact opposite position to that of Bellows. Unlike Bellows, World War I rehabilitation advocates saw the pension system as an ominous symbol of centralized governmental power, not the military hospital. Moreover, rehabilitators believed that a disabled soldier would find his autonomy and independence through institutionalization, not through pensions. Though Bellows and World War I rehabilitators differed in most respects, they did share a common goal: reducing long-term involvement of the government in the veteran's everyday life.

Rehabilitation hospitals such as Walter Reed and Letterman purported to accomplish this goal through medicine. By holding out the promise that disability could be cured, rehabilitators convinced themselves that once soldiers were discharged, little to no further federal assistance would be required.

FIVE

The Limb Lab
and the Engineering
of Manly Bodies

For most combatant nations in Europe, the severity of mu-
tilations wrought by the Great War was unprecedented and
unexpected. Four million German soldiers were wounded
between 1914 and 1918, with an estimated 67,000 of these
men losing one or more limbs.[1] Great Britain experienced
similar rates of injury, and by the end of the war 750,000
ex-servicemen were deemed "permanently disabled," with
41,000 losing an arm, a leg, or both.[2] To make matters worse,
European artificial limb makers could not keep up with the
demand for replacement limbs. While serving in France,
American orthopedist Robert Osgood estimated that for the
year 1915, French manufacturers were able to produce only
700 limbs for its 7,000 amputees.[3] In June 1918, one British
hospital reported that it had a list of 4,321 amputee soldiers
still waiting to receive prosthetic arms and legs.[4]

Upon entering the war, the War Department thought it
unlikely that the United States would ever find itself in such
a dire situation, since America had a reputation for being one
of the largest producers of artificial limbs worldwide.[5] The
boom years for US prosthetic manufacturing began during
the Civil War, when the Union incentivized limb production by giving its 35,000 amputee-veterans a "limb allowance" for the purchase of replacement limbs. The business
continued to thrive during the second industrial revolution,
when "the flywheels and pulleys of the new mills and facto-

8   J. E. Hanger limb makers working at Queen Mary's Hospital in Roehampton, England. Records of the Office of the Surgeon General (Army), Record Group 112; National Archives at College Park, College Park, MD.

ries severed arms and legs with alarming frequency . . . as did the wheels of railroad locomotives."[6] By the start of the Great War, there were approximately 200 artificial limb manufacturers in the United States, leading Great Britain to invite the largest American prosthetic companies, J. E. Hanger, J. F. Rowley, and A. A. Marks, to set up workshops at the main amputee center, Queen Mary's Hospital in Roehampton, to help the Allies scale up production.[7]

And yet, despite the country's obvious advantage, the Medical Section of the Council of National Defense (CND), along with the Army Office of the Surgeon General (OSG), created the Artificial Limb Laboratory, called the Limb Lab, in order to study "the artificial-limb problem."[8] Most strikingly, the OSG gave the Division of Orthopedics full control over the lab and ordered that the work be undertaken at Walter Reed General Hospital in conjunction with the Army Medical School. The decision to bring limb making under the authority of medical doctors ushered in a new era of prosthetic manufacturing that would have implications for the rest of the twentieth century. (The Limb Lab still exists today as an integral part of Walter Reed Medical Center.)[9] Before the Great War, artificial limb manufacturers were tradesmen. Some came to the practice as amputees themselves with a desire to make better limbs for their own use. Others came to it with backgrounds in clock and cabinet making. As businessmen owning their own shops, limb makers prior to World War I "generally dealt directly with the public and the amputee without mediation by a physician."[10]

If the United States had a steady supply of artificial limbs, what, then, was the "problem" that the CND and OSG hoped to solve through the creation of the Limb Lab? And why, out of all the various patients whom orthopedists treated did amputees receive so much focused attention and concern? According to statistics compiled during the war, more soldiers suffered from ankylosed joints than amputations.[11] Nevertheless, in popular and medical circles amputees provoked a more emotional reaction than did soldier-patients with joint contractures. The visual shock of dismemberment served as testimony to the brutality of war.[12] Yet while limblessness had been interpreted as a "badge of courage" in post bellum America, amputee soldiers became symbols of fear and waste during the Progressive Era.[13] More specifically, antipension Progressives worried that the amputee soldier would "prey" on an onlooker's charity and sympathy by displaying his missing limb, accentuating his injury with a pinned up sleeve or a peg leg.[14] The amputee who failed to hide his disability and instead used the obviousness of the physical deformity to gain money through charity or through government pensions represented everything that antipension Progressives wanted to change.

In a real sense, the so-called artificial limb problem of the First World War—much like the pension problem itself—stemmed from the Civil War. Providing $75 to buy an artificial leg (and also often paying for rail travel to the nearest limb manufacturing site) and $50 for an arm, the US government had spent $500,000 on replacement limbs for Union veterans by 1870, at least, in theory. Since there was no monitoring of how the money was spent, the government had no way of knowing where exactly the money was going. Because purchase of an artificial limb was not required by law, some veterans no doubt pocketed their limb allowance money, buying the cheapest leg on the market—a peg leg, for instance—while using the cash difference for other expenses.[15] By the time of the Great War, rumors abounded that veterans from both the Civil War and Spanish-American War used their limb money for buying liquor, an act of potentially scandalous proportions in a nation on the cusp of ratifying the 18th Amendment prohibiting the manufacture and sale of alcohol. Others who wished to have top-of-the-line prosthetic limbs—devices that could cost up to $120—used their limb allowance and supplemented it with their own income. Still others bought no replacement limb at all, preferring to adapt to their new physical state without any assistive device.

The CND, OSG, orthopedists, and other rehabilitation advocates opposed the laissez-faire nature of the Civil War limb allowance. The negative social and political connotations bound up with the aesthetics of limbless and peg-leg veterans played such a powerful role in the workings of the Limb Lab that even in the face of mounting evidence that some European amputee veterans preferred to adapt to their disability rather than wear cumbersome artificial limbs, the OSG forcefully mandated artificial limb wear, creating legislation that made it virtually impossible for US amputee soldiers to be discharged from military service without months of rehabilitation (under military observation) and a daily routine of artificial limb wear.[16] The Limb Lab's goal was to give every man, whether legless or armless, a "modern limb"—a limb that would make it possible for amputee soldiers to pass as normal, able-bodied citizens in the workplace and on the streets. By masking the disability through mandatory limb wear, rehabilitators aimed to delegitimize the disabled veterans' claim to federal assistance once rehabilitation was complete.

Speeding-up and Standardizing Limb Making

When the Limb Lab opened in September 1917 at the Army Medical School in Washington, it did not look much like a scientific laboratory. With limb-making equipment on order and not to be delivered until January 1918, the lab was little more than an empty office where orthopedic surgeon David Silver sat at a desk, focusing on paperwork. Silver could not immediately engage in hands-on experimentation because he had first to educate himself on the subject of the lab work. Before Joel Goldthwait and Elliot Brackett recruited their fellow Harvard Medical School graduate to become assistant director (second in command) of the Division of Orthopedic Surgery in early 1917, Silver worked as an orthopedic surgeon at both the Allegheny General Hospital and the Pittsburgh Hospital for Children.[17] Coming from the "great basin of the upper Ohio [River]" with its "huge manufacturing plants and its vast coal fields," he had his fair share of experience with fractures and wounds, injuries that he would later liken to wounds of warfare.[18] But upon entering war service, he had relatively little experience with amputations and artificial limbs.

As an orthopedic surgeon, Silver's qualifications were not unusual for his time. In the early twentieth century, general surgeons or industrial surgeons (physician-employees of specific manufacturing or coal mining businesses) performed far more amputations than orthopedists did. Orthopedists had little to offer the amputee, for medical care customarily ended once the patient's stump healed and was free of infection. The decision to acquire an artificial limb was left up to the individual patient who would then seek out the nearest commercial limb manufacturer. Up until the First World War, prosthetics was a business, not a medical subspecialty.

The creation of the Limb Lab would change all this, bringing prosthetic design and wear under medical control—specifically orthopedic surgery. The CND and the OSG established the Limb Lab so that the army could provide mass-produced, standardized artificial limbs to every amputee soldier. Standardization, the CND and OSG maintained, would "reduce the cost" of production and improve the "efficiency" of output.[19] In other words, the US federal government wanted artificial limb production to become more like its munitions industry. The system of standardization and the use of interchangeable parts—known globally as the "American System of Manufacturing"—began in US armories during the mid nineteenth century and eventually expanded during the late nineteenth century to include commercial industries that manufactured typewriters, sewing machines, and clocks.[20] Standardization and interchangeability became the defining features of modern productivity, a "technological religion, the supreme emblem of the achievements of inventiveness of American industry."[21]

But to military strategists and Progressive Era reformers, standardization promised more than mere efficiency: it was a way to impose a kind of cultural uniformity upon a disparate lot of soldier-citizens drafted into the war. Most soldiers who fought during the Civil War and Spanish-American War were volunteers who represented individual states in uniform and identity.[22] The First World War was different in that the federal government made its first attempt to raise a national army, drafting civilians through the Selective Service Act. By war's end, over 72 percent of the men who served during the Great War were conscripted. "Civilians came into the military from all walks of life," writes World War I historian Jennifer Keene, "and they brought a tremendous range of civilian attitudes and behaviors with them." Approximately 18 percent of all enlisted soldiers during the First World War were foreign born and 13 percent were black; a half million foreign-born troops serving in the military could not speak English.[23] Standardized uniforms, standardized physical training, and standardized rules of discipline—and standardized replacement limbs—would lead, the military and Progressive reformers argued, to a universal set of values and mores among the troops, bringing everyone from the white, college-educated officer to the Russian-born private into consensus. The standardization that went along with militarism would be a vehicle for Americanization.

The artificial limb industry resisted standardization as a mode of production, for its practitioners identified themselves as highly skilled craftsmen well into the twentieth century. The construction of a single wooden leg could take years to complete. By the time of the First World War, over 94 percent of American prosthetic firms used wood to make artificial limbs. To get a sense of the complexities of the process, consider the work that went into preparing the wood before it could even be put on a lathe. Prosthetists purchased wood bolts (a log of wood, standardized in shape for sale) and then "seasoned" it by storing it in an arid, dark, airtight shed until the wood was completely dry. The bolt needed to be moisture-free so that the eventual leg would be "stable," resistant to expansion and contraction when it was exposed to changing weather conditions or perspiration. Dried-out wood was also necessary for the "finishing" process, when a prosthetist used glue and other adhesives to cover the limb in rawhide. While a few of the larger manufacturers could artificially dry the wood through kilning, speeding up the seasoning process, the overwhelming majority of prosthetists let bolts of wood sit in their workshops for years until they were ready to use.[24]

The actual crafting process took a considerable amount of time as well as patience. Before a bolt of wood could be turned and smoothed on the outside and hollowed out on the inside to produce a "socket" (a smooth bucketlike depression into which the amputee would place his stump and through which he would control the movements of the limb), the prosthetist met with the patient, face-to-face, in order to measure and cast the stump. Once the prosthetist finished a rough version of the leg, he would see the patient again to test the fit. The prosthetist would rework the leg and socket numerous times until the fit was acceptable to both the patient and the limb maker.[25]

Silver's job as chief of the Limb Lab was to take artificial limb production in a new and modern direction. Two factors in reaching this goal worked in his favor. First, as an outsider to the trade, he had no commitment to maintaining the craftsmanlike image of limb making. Second, as an orthopedic surgeon, he had experience running hospital-based appliance shops, with mechanics working under him. By the time of the Great War, orthopedic surgeons had become accustomed to having brace-making shops with skilled metal and lathe workers to operate the machinery (leather sewing machines, upright drills, and grinders) to make orthopedic appliances.[26] In many ways, the Limb Lab was an extension of the orthopedic appliance shop, requiring Silver to exhibit more managerial than medical acumen.[27]

To create a standardized limb, Silver had to first establish a uniform set of rules of manufacturing in an industry where no such thing existed. Prosthetists disagreed passionately with one another about how limbs should be constructed, including how they should look, fit, and feel. While a majority of limb makers relied on wood, other manufacturers used fiber and steel. While some believed that the best artificial foot was made of rubber, others advocated wooden feet. While some thought that an articulated, movable ankle joint was better for ambulation, others preferred a solid ankle, with only the toes moving, believing that this type of mechanism helped the amputee to achieve a more lifelike gait. To get a sense of just how many different kinds of limbs were on the market, consider that by the turn of the twentieth century the US government had granted over 150 patents for artificial legs.[28]

Amidst these disputes and discord, Silver aimed to distill the commonalities among limb makers, looking for a model prosthesis that would require as little maintenance as possible and would be durable, functional, inexpensive, and adaptable enough to fit a wide range of male bodies. In order to arrive at this ideal, he conducted a massive, nationwide survey, sending out sample questionnaires to over a hundred limb manufacturers. He asked prosthetists questions about the best site for amputation, the most favorable time for fitting amputees with limbs, and the causes of stump ulcerations, a condition that kept amputees from being able to wear artificial limbs. Although the questions on the survey could have been answered in one or two sentences, most limb makers sent back the questionnaire with pages of additional typewritten answers, providing highly detailed and medically informed responses to Silver's inquiries.

While prosthetists cooperated in answering Silver's questionnaire, they voiced hostility toward the federal government, the medical profession, and the Limb Lab. After holding a vote at a meeting of the Association of Artificial Limb Manufacturers (AALM) in the fall of 1917, the industry's leaders drafted a letter to Silver and the Surgeon General's Office, outlining certain conditions to which the army had to agree before they would provide information and assistance to the Limb Lab. First, AALM prosthetists requested greater reimbursement fees for the artificial limbs supplied to US veterans, fees first established under the Civil War pension system. In the letter, prosthetists complained that the government's fee scale for limbs had remained unchanged since the 1870s. In addition to raising the prices of limbs issued by the government, the AALM insisted that the practice of awarding government contracts to one or several artificial limb companies, as other combatant countries were doing, be prohibited.[29] The AALM wanted to continue to foster the free-market practices upon which the field of prosthetics had been built since the Civil War.

While the OSG initially remained silent on the question of price scale, it eventually supported the AALM in its contention that government contracts should be avoided.[30] Silver wrote to Goldthwait, "if the Government were to make the legs . . . it would involve the question of Government ownership and interference with an industry which has been highly developed in America." Unlike the Canadian system, where government-controlled "limb factories" supplied one type of artificial appendage for all of the country's maimed, the Surgeon General's Office believed that such a system would falter in the face of the commercial forces that thrived in the American marketplace. Silver put it plainly when he predicted that American soldiers would automatically be "dissatisfied with a Government leg, no matter how excellent, due to the commercial houses [of artificial limb makers] trying to sell the men legs of their own manufacture."[31] Silver thus planned to come up with a blueprint for a standardized military-issue arm and leg that could be built and repaired in house at his own Limb Lab as well as off site in the nation's nongovernmental limb manufacturing firms.

The AALM prosthetists, however, wanted more than mere assurances of increased pay and a free marketplace; they wanted to assume responsibility for the Limb Lab entirely. J. F. Rowley, of the Rowley Artificial Limb Co., for one, offered to take over the lab, maintaining that he had far more expertise than Silver or any other orthopedic surgeon. A leg amputee himself, Rowley started his business during the 1880s, when, out of frustration with the prosthetic legs available on the market, he invented his own "Rowley Leg."[32] Rowley Artificial Limb Company became such a success that by 1917 it had manufacturing houses in Pittsburgh and Chicago. Moreover, Rowley claimed to produce over 500 limbs per month for the British hospital in Roehampton.

Like many other prosthetists in business at the time, Rowley used his own one-legged state to claim authority over able-bodied medical professionals.[33] He told Silver that being an amputee gave him a distinct advantage, for he could "test out every artificial leg ever marketed . . . in [the U.S.A.] and Great Britain."[34] William T. Carnes, an arm amputee and inventor of the highly popular "Carnes Arm," made similar claims to expertise on the basis of his own life experiences as a one-armed workingman.[35] Prosthetists who were not amputees themselves often hired amputee employees in order to demonstrate the legitimacy of their wares.

REASONS WHY
MARKS PATENT
ARTIFICIAL LIMBS
WITH
RUBBER HANDS & FEET
ARE THE BEST
TIGHT
ROPE
WALKING

THE
TRIVMPH
OF
ARTIFICIAL
LIMB
WEARING

BECAUSE

A.A.MARKS,   701 BROADWAY, NEW YORK.

9   A. A. Marks advertising pamphlet, circa 1900. Notice how the tight-rope walker, with an
     artificial leg, skillfully traverses a thin cable, one "foot" over the other. Courtesy of the
     Historical Medical Library of the College of Physicians of Philadelphia, Philadelphia, PA.

A. A. Marks Limb Company, for instance, sent one such employee, Frederick M. Voss, to Europe during the war in order to improve sales. In a letter to Silver, Voss wrote of his experiences, saying that "the people of the French and English embassies found it perfectly marvelous that I, with only a seven inch stump, am not only able to walk perfectly natural with my artificial limb, but can also run, jump, dance, kneel down, spin around . . . and do military exercise . . . on the artificial limb."[36] Artificial-limb makers and their amputee-employees used performance as a way to attract potential buyers to their products, entertaining the eye with fantastic stunts of physical prowess, telling stories of bodily triumph over disability and defects.[37]

The OSG and the Division of Orthopedic Surgery had no intention of giving prosthetists control over the Limb Lab, for prosthetists had resisted the primary aim of the lab—namely, standardization—from the outset. Limb manufacturer E. H. Erickson wrote to Silver, telling him that there were "no fixed rules that [could] be followed" in the making of artificial limbs and that "fitting is mostly hand work," requiring the patient to visit the prosthetists at his shop personally.[38] The president of the Doerflinger Artificial Limb Company in Milwaukee maintained that artificial limbs were "almost as different as finger prints."[39] At the late date of 1936, J. E. Hanger made a similar observation as part of an advertising brochure:

> The making and fitting of an artificial limb requires, in every stage, the closest attention to the distinctive anatomic characteristics of each individual in order to ensure that it conforms, with meticulous precision, to the shape and any peculiarities of the stump.

No two cases are alike; not only the type of limb but also the method of attachment which suits one patient may be most unsuitable for another . . . as stump condition, age, occupation, sex, physique and general health all need careful consideration. . . . We are satisfied that mass-production methods and attempts to standardize limbs in different sizes—even if they run into the thousands—are wrong in practice, definitely harmful to the unfortunate wearers, and much to be deplored. Making an artificial limb is considerably more than an engineering problem.[40]

But the Limb Lab disagreed and wanted to do precisely what Hanger and his colleagues thought impossible: turn artificial limb wear into a social and material engineering problem.

Silver blamed the prosthetists' resistance to standardization on their traditional, woodcraftsman's mindset. In one letter, Silver attributed the prosthetists' behavior to their lack of "professionalism." The "current art of artificial limb manufacture," he wrote, was based on knowledge "transmitted in the ancient manner—by word of mouth," not by scientific measures and professional modes of communication through text.[41] Prosthetist Charles Doerflinger took the exact opposite stance. "The legitimate fitting of amputations with artificial substitutes," he wrote Silver, "is a profession, and not a manufacturing position." In other words, Doerflinger believed that limb makers were professionals by virtue of their artisanship, a practice based on esoteric knowledge about mechanics, fabrication, and the human body. To him, standardization threatened his professional standing by turning limb making into an assembly-line mode of production. Not wanting his profession to be modernized or his colleagues to be treated as blue-collar laborers under a physician's watchful eye, Doerflinger ultimately maintained that "to produce artificial limbs by the ordinary factory methods with the speeding up process [would] not . . . benefit the patient."[42]

The Liberty Limb

Sensing overwhelming opposition to standardization among prosthetists, Silver distanced himself from many of the industry's major owners by December 1917 and instead relied on a few lesser-known limb makers for advice on the use of alternative materials and methods of production.[43] Early in his post as chief of the Limb Lab, Silver had his eye on a new process of limb making that used compressed wood fiber or pulp (as opposed to solid wood) to construct artificial legs. At the time, very few companies had even attempted to use fiber for limb construction.[44]

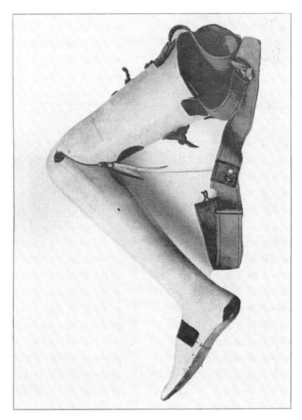

10 An example of an E-Z-Leg for an above-knee amputation. Reproduced from the United States Army Surgeon General's Office, *The Medical Department of the United States Army in the World War* 11, pt. 1 (Washington, DC: GPO, 1927), 741.

According to a 1917 survey conducted by the US Department of Agriculture Forest Service, such material had not been sufficiently established as a reliable means of producing sound limbs. Out of 151 limb firms surveyed, only six companies used vulcanized fiber. The Department of Agriculture report concluded that the use of fiber was purely "experimental."[45]

One little-known business, however, the E-Z-Fit Artificial Limb Company located in New York City, professed to have perfected the use of vulcanized fiber in limb making. In December 1917, E-Z-Fit president, Albert G. Follett, wrote Silver, boasting of his lightweight legs made of "fiber shells."[46] Follett sent samples of his E-Z-Legs to Silver's lab.

Silver took a sample limb to Walter Reed General Hospital to have it tested on amputee patients. There Silver and his orthopedic colleagues studied the construction and mechanics of the E-Z-Leg. Familiar with the properties of vulcanized fiber, Silver knew that the two greatest difficulties facing the use of such material were durability and the difficulty of molding it into shape.[47] But after several patient trials, the Limb Lab researchers concluded that the E-Z-Fit Artificial Limb Company had produced a durable, modern-looking limb that closely resembled the shape of a real-life human leg.

From the Limb Lab's perspective, the E-Z-Leg satisfied almost every need particular to the war. Unlike the more traditional manufacturers, Follett promised a modern method of production. The leg was manufactured in sections. Thigh pieces were made separately from shin shells, and shin shells separately from wooden feet. Moreover, such a method of production lent itself easily to the standardization of sizes. Follett assured Silver the E-Z-Leg could be "manufactured in several standard sizes with reference to circumference."[48]

Because E-Z-Legs could be produced speedily and delivered in sections rather than whole pieces, shipping legs in a cost-efficient manner to American wounded soldiers overseas became a possibility. In January 1918, only three months after the lab opened, Silver made arrangements to have the first batch of E-Z-Legs sent to Goldthwait in France.[49] Silver saw such an early deployment of these limbs as a triumph in modern medicine as well as in artificial limb production. Early limb wear, he believed, provided both physiological and psychological benefits to amputees. While artificial limb makers customarily preferred to wait until the final shape of the patient's stump had been achieved (a process that could take months, if not years) before fitting an amputee with a prosthesis, Silver found this tactic to be "too passive." Instead he encouraged an aggressive program of massage, exercise, and calisthenics so that the stump could heal more speedily. Once surgical healing had been attained, Silver maintained that "toughening" the stump by "pounding it on a firm surface" should be "vigorously pursued."[50] Such "pressure exercises," he concluded, would decrease the chance of reinfection by promoting better circulation through weight bearing. He also argued that early weight bearing helped the stump to "shrink" (i.e., a reduction in edema after surgery), permitting the amputated limb to "attain its final form," thus making it possible to fit the patient with a permanent limb.[51]

Getting patients to be active participants in their care as early as possible fell in line with the overall goals of rehabilitation itself. Silver's own use of the descriptor "passive" to portray care of amputees prior to World War I in contrast to his "active" regime of stump pounding and early prosthetic wear demonstrates that even something as seemingly value-neutral as stump care was rife with normative meaning. Just as the manufacture of limbs was to go through a speeding-up process, so too were amputee patients expected to heal quickly. Accordingly, Silver measured success in amputee care by speed of recovery. "The aims," wrote Silver, "are as follows: (1) to *hasten* shrinkage of the stump, (2) to apply a prosthesis at the *earliest* possible moment, (3) to get the patient walking *as soon as possible,* (4) to avoid the use of crutches, or *shorten the time* they may be needed and (5) to *hasten* [the soldier's] return to productive economic employment" (emphasis added).[52] Silver also advocated speedy healing and early limb wear so that ship transports of wounded men back to the United States could be made easier. Instead of utilizing able-bodied men to carry stretchers of amputee soldiers on and off ships, amputees with provisional limbs could board and exit independently.[53]

The E-Z-Leg pleased those who held the army purse strings, as well. Because he engaged in mass production, using materials that did not require years of preparation and that cost a fraction of the price paid for solid wood bolts, Follett was able to offer the army a relatively inexpensive product. Whereas a solid wooden leg from a company such as Hanger would have cost $200, Follett charged the US Army only $20 per leg.[54] The price of an E-Z-Leg not only cost less than most other legs on the market at the time but also less than what the government paid for limb allowances to Civil War amputee veterans.

All in all, the Limb Lab believed that the E-Z-Leg provided freedom and autonomy, both to the amputee and to everyone involved in the effort to provide wounded soldiers with replacement limbs. By the summer of 1918, the Limb Lab had dubbed the E-Z-Leg the Liberty Leg.[55] With the Liberty Leg, amputee soldiers could don replacement limbs like ready-made clothing sold at large department stores. The E-Z-Leg's standardized thigh, shin, and foot units were all interchangeable, making it possible to build legs using combinations of parts, simply by driving wooden dowels into the articulating joints.[56] Moreover, since the vulcanized fiber parts could be easily sawed off, the wearer could adapt the length of the limb to his own liking. In other words, the E-Z-Leg promised the amputee and the military complete autonomy over the care and construction of his leg. Wounded men could be fitted immediately after surgery, overseas, without the need of a prosthetist or a limb shop.

The Problem of the Peg Leg

At the same time that Silver ordered large numbers of E-Z-Legs sent to base hospitals in France, the Bureau of Artificial Limbs at the American

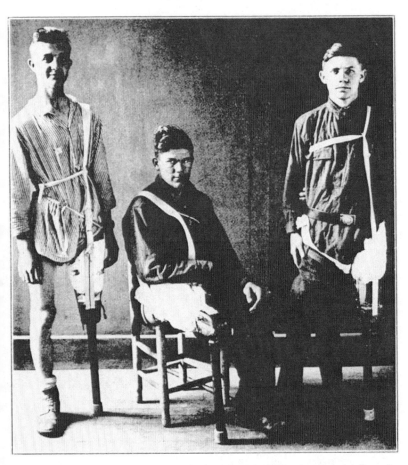

11 Paris Bureau Red Cross patients wearing peg legs. The United States Army Surgeon General's
Office, *The Medical Department of the United States Army in the World War* 11, pt. 1 (Washington, DC: GPO, 1927), 704.

Red Cross in Paris, a facility that had been established by the United
States in the spring of 1918, chose to fit its lower limb amputee soldiers
with traditional peg legs, devices used during most of America's previous
wars. While the physicians working at the American Red Cross employed
logic similar to Silver's concerning the physiological and psychological
benefits of early weight bearing, they disagreed about the best design for
limbs. Physicians at the Paris bureau argued that prosthetic appliances
needed to be "sufficiently simple" in design—straightforward enough
that they could construct the temporary limbs themselves. Peg legs required only the most basic of materials, supplies that could be found at

111

any World War I base hospital equipped for orthopedic cases. For upper thigh amputations, bureau physicians used plaster of Paris to make a stump socket, steel frames to hold the socket, and a strong wooden peg or crutch, with a rubber tip, to complete the leg. The design was rudimentary, but it worked, making it possible for "patients [to] get out of bed and walk without other support [i.e., crutches] very shortly after amputation."[57]

Moreover, because bureau physicians were able to observe the war wounded first hand, they saw no need for elaborately designed limbs. Many of the amputees treated overseas suffered from infected stump wounds and ulcers, making limb wear impossible. "It was realized from the beginning," wrote one bureau physician, "that the percentage of amputation cases in the American Expeditionary Forces to whom [limbs] could be [given] would be small." The initial evaluation of the situation proved to be correct, for, as the Surgeon General's Office later reported, only 20 percent of all lower extremity amputee cases abroad had healthy enough stumps for limb wear.[58]

The example of the Paris bureau demonstrates that Silver had other and indeed cheaper artificial limb options before him. But that raises the question: Why did Silver not choose the age-old peg leg, a limb that was cost effective, did not require off-site manufacture or shipment overseas, and that had stood the test of time in terms of durability? Answering the question requires a careful consideration of Silver's convictions concerning the aesthetics of amputees and artificial devices and the role that American medicine, culture, and politics played in shaping his sensibilities. When describing the E-Z-Leg to his friend and colleague overseas, Silver wrote that it was a "*real* artificial leg, shaped and finished like any other" human leg (emphasis added).[59] At other times, Silver heralded the E-Z-Leg as a "modern artificial limb."[60] In order to convey what he meant by "real" and "modern," Silver frequently juxtaposed the E-Z-Leg with the traditional peg leg. As he told Goldthwait, the E-Z-Leg was visually "more pleasing to [a] man than a peg."[61]

For Silver, artificial legs were more than mere tools for standing and support—they were symbols, material entities that embodied sociopolitical fears and hopes, artificial devices embedded with stories from the past and visions for the future. Among middle-class Americans, peg legs tended to conjure up the frightening images of the bloodthirsty fictional characters, such as Captain Ahab from Herman Melville's *Moby Dick*. In a similar vein, hook-arms became theatrical tropes, material representations of immorality and deceitfulness.[62] To cite but one example: in the 1908 film *The Thieving Hand*, an upstanding one-armed pencil vender turns into an unlawful pickpocket after buying a new hook-arm. Not until the vendor gets rid of his artificial arm does he return to his former, honest self.[63]

In addition to these general feelings of distrust toward the half-machine/half-human body, US Progressives and fiscal conservatives of the early twentieth century associated the peg with the hobbling remnants of an ineffective Civil War system that produced pensioners and veterans' homes. US reformers also worried about a postwar influx of unemployed peg-leg soldiers into the cities, begging for money and adding to an already growing population of "undesirables" in urban areas. In short, the peg leg was an easily identifiable cultural symbol of want, need, and charity that Silver and others like him wanted to stamp out.

The E-Z-Leg thus provided Silver with a solution to the societal fear that World War I ex-servicemen would expect charitable handouts as their predecessors had. In a report featuring the E-Z-Leg in the *New York Times*, Major P. B. Magnuson, of the Medical Reserve Corps, proclaimed that "with this [leg] Uncle Sam will be able to avoid conditions which followed the Civil War, when so many one-legged soldiers hopped along the streets on crutches."[64] Mimicking the shape and look of an actual, living limb, the E-Z-Leg had no culturally shared symbolic meaning. If anything, the E-Z-Leg was to serve as a cultural foil, a device that everyday people would mistake for a real leg. Ultimately, the hope behind the E-Z-Leg was that it would prevent veterans from misusing their artificial limbs as a means to garner unwarranted sympathy or alms.

Silver opposed peg legs to such a degree that he, along with the Division of Orthopedic Surgery, convinced the army surgeon general to discontinue the Civil War policy that allowed veterans to purchase peg legs independently. He urged the OSG to establish stricter surveillance over the limb allowance program, not permitting any amputee soldiers to leave military duty without an E-Z-Leg or another like it.[65] This was a monumental step for the army to take. In Germany and England, for instance, amputee soldiers were dismissed almost immediately from military service (unless they were healthy enough to be sent back to the front lines) and referred to private rehabilitation programs.[66] But in the name of replacing the pension system with a full-fledged program in rehabilitation, Silver persuaded the US Army to funnel all amputee soldiers to Walter Reed General Hospital, where they would be observed, fitted for limbs, and maximally rehabilitated before they would be released from service, a regimen that could last anywhere from eight weeks to a year.[67]

In addition to his concerns about the cultural meaning of the peg leg, Silver worried that his amputee patients would never wear such a device.

"I do not feel," he wrote to Goldthwait in December 1917, "that the peg leg is going to be acceptable to our men."[68] At this point in his tenure as lab director, Silver had already received word from rehabilitation centers in Germany and Canada that many amputees refused to wear artificial limbs. According to a 1917 German report, 310 out of 356 artificial arms distributed in the Rhineland had been discarded by their owners.[69] Similar complaints were coming from Canadian rehabilitation workers, who claimed that they were "experiencing great difficulties in inducing [amputees] to wear their artificial limbs."[70] Upon receiving this news, Silver made it a goal to surpass all other belligerent countries in patient compliance rates. One way of accomplishing this task, he believed, was through the E-Z-Leg. A lifelike leg, rather than a peg, he told Goldthwait, would be better for a soldier's morale, "especially at first when [the soldier] is likely to be depressed by his loss." He hoped to gain the trust and enthusiasm of US amputee soldiers by providing them with a modern-looking limb that would, in his words, "cheer" them.[71]

Putting Standardization to the Test: The Limb Lab Moves to Walter Reed Hospital

While standardization of artificial legs seemed like a good idea in theory, problems arose when the limbs were put into use. In the summer of 1918, after the first wave of US amputees returned home for rehabilitative care, the Limb Lab moved to Walter Reed Hospital in order to "better coordinate the experimental and the clinical parts of the work."[72] Before that fall, the Limb Lab supplied prostheses to over 300 amputees.

Although the historical record concerning the amputee-patient's perspective of the Limb Lab is scant, official military sources indicate that many of the soldiers who were given an E-Z-Leg found the appliance cumbersome and early weight bearing painful. Following stump pounding exercises, "patients usually complained of discomfort," read one report. Another report stated that when amputees were forced to wear artificial limbs soon after surgery, they often "expressed gratitude when the artificial limb [was] removed."[73] In addition to being painful, the E-Z-Leg was unreliable. The Bureau of War Risk Insurance's chief medical advisor estimated that 50 percent of the E-Z-Legs that he had seen by May 1919 "were broken beyond repair." He wrote to Silver and his team urging them to construct limbs "of a more durable character."[74] In addition to complaints about the quality of the limb, patients who wore the E-Z-Leg objected to the fact that the leg did not easily adjust to stump shrinkage or difference in stump size. One group of patients reported that they needed to wear "sometimes as many as six or seven" stump socks for the limb to fit properly.[75]

That Silver and his colleagues did not see such complaints coming when they first set out to create a standardized, one-size-fits-all limb, is a testament to their blind willingness to favor an abstract ideal of productivity over the reality of patient individuality and difference. Even their colleagues in Great Britain saw the troubles that a standardized limb could cause. As one army colonel at Roehampton put it: "A leg is not a suit of clothes—it must become a . . . lasting personal adjunct. . . . questions of temperament, nerves, and general health govern each case." "No two men's legs," he concluded, "are the same and standardization of the actual limb is therefore impossible."[76]

In order to reconcile their goal of standardization with the reality that draftees came from many different races and ethnicities, Silver and his colleagues used the white male body as the uniform ideal in amputation care and limb design. Although he knew that Walter Reed Hospital would be treating African-American soldiers alongside white soldiers, Silver nevertheless came up with the specs for the E-Z-Leg using anthropometric leg measurements compiled by army surgeons in segregated, all-white training camps.[77]

For all of Silver's boasting about the authenticity of the E-Z-Leg, the fact of the matter was that the limb looked white, in size and color (see fig. 10). Silver followed the golden rule in limb production at the time: make replacement limbs with white wearers in mind.[78] "After the leg has been shaped to as near a duplicate of the wearers' natural limb," read one army limb-making instruction manual, it should be "covered in rawhide. . . . [with] the application of several coats of white . . . colored enamel."[79] The Limb Lab did not attempt to match the color of the E-Z-Leg to that of its wearer, for to do so would have undermined the goals of uniformity and standardization.

In order to overcome the shortcomings of the E-Z-Leg and to insure that, despite its faults, patients would wear it, Silver and the Limb Lab employed various tactics of persuasion to insure that amputee-patients were invested in their artificial limb, much as they would be in a real limb. The Limb Lab, for example, became a place where amputees would not only be fitted for limbs but also be educated on how to construct and repair the limbs themselves. According to orthopedist Leo Mayer, Austria engaged in a similar practice, having its amputee soldiers serve at least four weeks in the artificial limbs department in order to learn how to repair their own limbs.[80] User repair, whether through exercise, curative workshop, or limb modifications, was an ideology inherent to the rehabilitation project, for it represented the pinnacle of autonomy and independence.

A further means of persuading amputees to wear their artificial limbs was to make them believe that their prosthesis was essential to work and leisure. Take, for example, the Limb Lab's "utility arms," designed and prescribed for upper limb amputees. The idea for a "utility arm" came from the Chicago-based Dorrance Artificial Limb Company. Advertised as "The Arm that Arms You for Work or Play," the Dorrance arm appealed to Silver and his team because of its functionality: it acted more

12 Example of amputee patient using one of the Limb Lab's "utility arms" with a welding attachment plugged into the arm. The United States Army Surgeon General's Office, *The Medical Department of the United States Army in the World War* 13 (Washington, DC: GPO, 1927), 338.

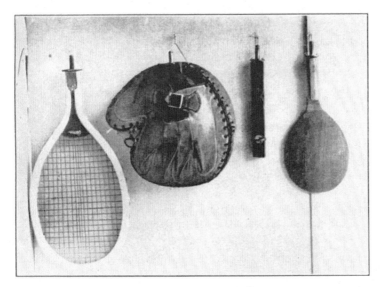

13 The Limb Lab invented sporting attachments to plug into the standardized "utility arm" issued to arm amputees. The United States Army Surgeon General's Office, *The Medical Department of the United States Army in the World War* 11, pt. 1 (Washington, DC: GPO, 1927), 747.

as a tool than a real human arm and hand.[81] With the Dorrance arm, an amputee did not need to carry around traditional tools of work, because a work tool could be directly attached to the artificial limb itself. Like the Dorrance arm, there were multiple hand attachments for the army's provisional arm, interchangeable accessories that could be plugged into the forearm. A steel, nickel-plated hook was the most common "hand" given to arm amputees. Yet depending on a soldier's line of work, the hook could be replaced with a clamp, a sander, or any other form of adaptable specialized tool. An amputee thus became distinguishable by the tools attached to his arm, making arm prosthetics "markers of class identity." In the words of historian Heather Perry, "The prostheses firmly bound the worker to his profession, and by extension, to his class: a veteran with a specially crafted arm or set of working hands could not easily change jobs."[82]

To make the utility arm even more appealing, the Limb Lab invented specialized "play-time" attachments. Organized sport, Walter Reed surgeons believed, was a good way to coax amputee soldiers into wearing their artificial limbs—it took the patient's mind off his disability and redirected it toward the "competitive spirit" of sport.[83] Silver's colleague,

Dr. Carl Yount, took a particular interest in designing and developing such sporting accessories, inventing adaptive baseball mitts, tennis rackets, and ping-pong paddles, all of which mounted directly into the artificial arm.

Interestingly, unlike the attachments created for work, Silver and his team developed sport accessories that were decidedly middle-class in orientation. According to historians who study the relationship between class and leisure time, it is unlikely that the immigrant or working class patients of Walter Reed had much experience playing tennis, ping-pong, or baseball. Working-class men participated in boxing matches and frequented saloons more than they engaged in organized sport.[84] In the Limb Lab's effort toward standardization, it was white middle-class men who set the standard. The expectation, therefore, was that all nonwhite injured soldiers could be transformed into middle-class white men, in work, play, and in the "flesh."

*   *   *

Eight months after the armistice, virtually all of the US amputation cases— an estimated 4,400 in all—had already returned home and been fitted with artificial limbs. Despite Silver's strenuous efforts toward achieving a high rate of patient compliance, by the early 1920s the OSG published reports demonstrating that US rates of limb wear were no better than Europe's. For many leg amputees, the E-Z-Leg proved to be unreliable and uncomfortable. The utility arm did not fare much better. According to a 1927 report published by the OSG, "approximately 60 percent of individuals who had suffered the loss of a single arm [did] not find existing prostheses sufficiently useful to compensate for the inconvenience of wearing them."[85] Like their European and Canadian counterparts, many discharged US servicemen who had lost one arm found it easier to function in everyday life and at work by adapting to their disability (mainly by compensating with the other, intact arm) rather than wearing a replacement limb.

Yet while the individual histories of the E-Z-Leg and the utility arm were short-lived, the existence of the Limb Lab was not. Indeed, the creation of the Limb Lab brought with it significant changes in amputation care for the entire century to come. First and foremost, the process of measuring an amputee's stump, as well as the construction and fitting of limbs, all within a manufacturing house, became a thing of the past. After World War I, prosthetic construction and care increasingly became part of the growing modern hospital, where surgeons had direct contact with, and eventually control over, the making of artificial limbs (today, for instance, you need a prescription to get a prosthesis). Prosthetic shops were set up in army general hospitals and remained there, becoming a permanent fixture in the eventual hospital system of the Veterans Administration a decade later.[86]

The types of limbs produced during this time period changed as well. The peg leg became virtually obsolete (except for temporary in-hospital use) and replaced in later wars with legs that duplicated not only the shape but also the actions of a real leg.[87] Having the prosthetic shop in the hospital made it easier for prosthetists and surgeons to test new materials and fits, using the in-house population of amputees as ready-made experimental subjects. While the fiber E-Z-Leg proved to be insufficient in terms of durability, the army continued to experiment with similar kinds of lightweight, flexible materials, paving the way for the use of early plastics in artificial limb production during World War II.

But a historical analysis of the World War I Limb Lab reveals changes beyond those that occurred within the field of medicine or amputation care. By looking at the aesthetic and ideological assumptions that drove Silver and his team both in their decision-making processes and their hands-on work, one can see that they emerged out of a cultural and political fixation of the time. Because of various sociopolitical concerns bearing down on him, Silver insisted on a "modern" limb with a natural-looking foot, not a leg that looked like a crutch. Above all, the negative symbols of the Civil War–era peg leg or empty sleeve, two images associated with an old system of politics and of economic waste, were too potent to deny.

Propaganda and Patient Protest

The US government fought the Great War as much through propaganda as actual armament. Mobilizing prowar public opinion became crucial for America's military involvement, especially since President Woodrow Wilson won reelection in 1916 on a platform espousing wartime neutrality. Wilson marshaled an unprecedented use of federal power to build home front consensus. Within weeks of declaring war on Germany, he established the Committee on Public Information (CPI), an independent governmental agency that deployed prowar messages through posters, film, and speakers known as the "Four Minute Men," who would advertise the attributes of war while simultaneously raising the specter of fear concerning the German "Hun." To help bolster the CPI's prowar message, Wilson signed the Espionage Act in June 1917, making it a crime for American citizens to express antiwar sentiments.[1] Those in charge of the institution of military rehabilitation programs played a crucial part in strengthening prowar sentiment as well, using propaganda to convince a skeptical public that one of the frightening aspects of war—namely, maimed soldiers—could be overcome and solved through a specific vision of social and bodily engineering.

During and immediately following the war, the OSG sponsored and produced more than forty-two hospital magazines, including its premier journal *Carry On: A Magazine on the Reconstruction of Disabled Soldiers and Sailors.*[2] Contributors to the effort included a wide spectrum of medical

14 From the Walter Reed Hospital magazine, *The Come-Back*, this image with its caption is an example of how rehabilitation officials used satire to undermine the pension system. The caption reads: "This ladies and gents is private Jenks who lives at present at Walter Reed. The ladies and visitors make a fuss over him. And he the big stiff [*sic*] thinks the worlds owes a hero like him something. Guys like him who think the future will be one grand handout ought to be handed one like this. Then we could tell at a glance that he is too strong to work." *The Come-Back*, February 12, 1919.

and nonmedical politicians and reformers who supported passage of the WRIA, believing it to be a more enlightened form of veteran welfare than what had come in the past. Theodore Roosevelt, a leading advocate of heightened war preparedness and "national greatness," regularly wrote for *Carry On*. Joining him were other prominent American political figures such as American Federation of Labor president Samuel Gompers as well as Hull House affiliate Eleanor Clarke Slagle.

15 An image for the Walter Reed magazine convincing disabled soldiers that it was their patriotic duty to undertake rehabilitation and be grateful for it. *The Come-Back,* May 7, 1919.

Rehabilitation propaganda served many purposes. First and foremost, this literature aimed to convince disabled soldiers that rehabilitation was the best form of compensation for their injuries and that they should be grateful for the medical and vocational training that the US government offered them.[3] Rehabilitation hospital magazines, in other words, intended to persuade injured soldiers that it was unnecessary, unseemly, unpatriotic, and, at worst, unmanly to receive military pensions as their forefathers had. In image and text, rehabilitation propaganda frequently used the Civil War pension system as a counterpoint to orthopedic reconstruction. Judge Julian Mack, the framer of the WRIA, told readers of *Carry On* that "As the 'cripple' is passing, so [too] is the 'pensioner.'" Through rehabilitation, Mack maintained, the pensioner, "will become as obsolete as the old soldiers' home, and the other institutions and practices that world progress is leaving in its wake."[4]

The second goal of rehabilitation propaganda was to convince family, friends, and the entire American public that disabled soldiers would benefit more from a program of hospital-based reconstruction than from returning home and receiving care and financial help from their local communities.[5] Speaking directly to the families of maimed soldiers, poet and art critic Gelett Burgess wrote: "Perhaps you are one who thinks that a grateful government, after [the soldier] has given so much in its defense, should provide for him a life of perpetual idleness." "No doubt you are willing to support him," Burgess continued, "thinking he has done his share." "But of all kindnesses," he warned "this would be most destructive."[6] In addition to persuading family and friends that hospitalization was best, rehabilitation officials felt they had to reform twentieth-century business attitudes toward the disabled, convincing would-be employers—who were still adjusting to the new workmen's compensation laws—that armless and legless men would not be liabilities and could be just as productive and efficient as able-bodied men in the workplace.[7]

Although rehabilitation propaganda focused on the grim reality of engaging in combat—on soldiers left maimed and deformed from the destruction of war—all stories, according to *Carry On* editor Dr. Casey Wood, were to be "distinctly cheerful and optimistic [in] tone."[8] In this way, rehabilitation propaganda worked to maintain political support for both the passage of the WRIA and the war itself. If rehabilitation lived up to its promise of curing disabilities through six weeks of physical therapy and vocational training, then, the thinking went, the human and economic cost of war would be greatly minimized and made less visible to a potentially skeptical public. Rehabilitation officials knew that if they did not garner support for their cause, the voting public could bring an end to their effort, with an outcome similar to the Civil War pension system. Likewise, the Wilson administration hoped that the rehabilitation program would both stave off antiwar sentiments and quickly satisfy the needs of the new generation of disabled veterans while relieving the federal government of the need to make a long-term financial commitment. An idealized notion of rehabilitation is still used today both to bolster persistent militarism and deflect criticism about the aftermath of war.

But while the OSG was busy painting physical reconstruction in a positive light, a different branch of the US Army, the Inspector General's Office (OIG), was inundated with complaints coming from soldier-patients whose day-to-day experiences were anything but cheerful.[9]

Aggrieved soldier-patients expressed their dissatisfaction by writing letters to their friends and family, who would then lodge official complaints
with the OIG for investigation. Frequent complaints included the quality
and quantity of hospital food, poor sanitation, and delays in pay—conditions that could be ostensibly rectified from within the military bureaucracy. The more challenging complaints were those arising from bigotry
and authoritarianism. Since rehabilitation hospitals were desegregated,
some white soldiers conveyed outrage that they had to share quarters
with black soldiers, while others criticized the poor treatment that their
fellow black patients received. Still other soldier-patients complained
of unjust methods of discipline and punishment; patients who persis-
tently disagreed with their commanding physicians ran the risk of being
charged with insubordination, a punishment that could, at its worse, lead
to time spent in "the cage," that is, the hospital prison. In short, airing
complaints through letter writing was a very risky form of protest.[10]

In telling a story that juxtaposes propaganda with patient experiences,
it is tempting to cast the OSG officials and medical doctors as villains and
their patients as heroes who resisted military and medical domination.
But soldier-patients were not of one mind. Some extolled their medical
care and volunteered to run hospital-based newspapers, printed materials
that read very much like *Carry On* in message and tone. (For example, a
disabled captain recuperating at Walter Reed created the illustration in
fig. 14 above). Another group of Baltimore soldier-patients created a Fifty-
Fifty League, the constitution of which promised to "support existing
authority and make it increasingly effective" through self-policing.[11] By
the same token, not all the medical professionals involved with the rehabilitation effort thought or acted alike. Some physical therapists came to
their patient's defense, while others did not. Some surgeons were keenly
aware of opportunities for their own professional advancement; others
believed in the cause of the disabled veteran and worked tirelessly for the
betterment of their patients.

Nevertheless, when analyzed side-by-side, OSG propaganda and the
OIG reports reveal two very different views of physical rehabilitation—
one based on ideals, the other on lived experience. At stake was the question of what the nation owed its citizen-soldiers who were drafted into
war and became permanently disabled as a result of it. Most injured soldiers found the new social contract of the WRIA and its heavy reliance
on rehabilitation lacking. Many wanted more than the federal government was offering—a just system of treatment, greater medical attention,
extended hospitalization, better financial compensation, and improved
benefits after discharge. Disabled veterans desired many of the same things

that their treating physicians and therapists wanted, but for all its assurances and optimism, medical rehabilitation often fell woefully short of its promise.

Ideology Meets Reality

Colonel Willard F. Truby, medical commander of Walter Reed Hospital, began hearing rumblings of patient dissatisfaction almost as soon as the first wave of injured soldiers arrived for treatment in the spring of 1918. The first complaint came from Cadet William Dearing Davis, a well connected "society fellow" from Chicago, who was recovering from knee surgery that he underwent after an aviation accident in Italy. Davis told a commanding officer, who was also a family friend, that the treatment that he and his fellow patients received "made his blood boil."[12] None of his nineteen ward mates, he claimed, had been paid in eight months. Their street uniforms had been taken away to prevent them from going AWOL.[13] Men with missing arms and legs were given only makeshift "improvised stumps" that easily broke down. And one amputee, Corporal Leonard E. Jackson, suffered continual pain in his affected leg because the surgeons who performed the amputation overseas left a portion of a Hagedorn needle, used for suturing, in his leg.[14]

After conducting interviews with nineteen soldier-patients and five Walter Reed medical officers, Major David Lewis of the OIG concluded that Davis's grievances "were those of a nervous young man" and thus "did not require further consideration."[15] Aside from Jackson and Maurice Bekaert—an arm amputee who insisted that an artificial limb would "do him no good"—the other seventeen soldier-patients reported that they were satisfied with their care: the food was good, they said, and the hospital provided a "good place to sleep." Davis's complaints appeared to be unsubstantiated and his case weakened by the fact that Colonel Truby found him easily "excitable" and to have "a rather selfish attitude."[16]

Men like Davis did not fit the rehabilitation officials' idealized vision of a brave, selfless, hard-working injured soldier who would valiantly overcome his disability without complaint.[17] Injured soldiers were supposed to attack their disabled bodies, to fight them, much as they had fought the enemy on the front lines. "[We] are trying to makes these boys want to fight harder than ever," one home-front rehabilitation official reported, "but now they are fighting a new battle, [a] long, hard fight for themselves."[18] To refuse rehabilitation was tantamount to treason. The hospital magazine *Come-Back* made this message clear, arguing that "the

Hun" was not the only enemy to threaten the security and survival of the nation. Individual character flaws—such as laziness, selfishness, grouchiness, and discourteousness—on the part of recuperating soldiers would jeopardize the new regime of rehabilitation, making it a policy failure rather than a success.

The rehabilitation propaganda literature of World War I readily assumed, and also wished its readers to assume, that disabled soldiers undergoing rehabilitation had sustained their wounds in battle. At times, this supposition was made clear in the oft-repeated slogan that disabled soldiers needed to "go over the top," a phrase used on the front lines to refer to courageous soldiers who leaped out of the trenches and into en-

16 An image depicting the multiple enemies within, character flaws that World War I disabled soldiers needed to overcome in order for rehabilitation to be successful. *The Come-Back*, May 7, 1919.

17 Amputee soldiers pictured in a duel to demonstrate that disabled soldiers-patients still possessed manly courage and bravery. *Carry On* 1 (July 1919): 21.

emy fire. Images of disabled soldiers engaging in simulated battles further encouraged readers to presume that the photographed subjects had been maimed in the line of fire. The photograph "Beware of the One-Armed Man" aimed to prove that rehabilitated disabled men did not cower in the corner, but instead engaged in hand-to-hand combat.

Whether in the boxing ring or on the battlefield, face-to-face combat was thought to be the purest demonstration of courage and physical ability, the pinnacle of manliness in civilian and military contexts during the early twentieth century.[19] Indeed, the military measured a disabled soldier's worth by the location of his wounds, assigning more honor to a soldier who had frontal injuries because, it was assumed, he was hit when running *toward* the enemy.[20] "When men run away in front of the enemy," General John J. Pershing told his division commanders on the battlefield, "officers should take summary action to stop it, even to the point of shooting men down who are caught in such disgraceful conduct."[21] A rear injury was thus a sign of cowardly behavior and a weakened state of manliness, for it indicated that the soldier took the hit while running away from his opponent.

Yet in reality, many of the disabled soldiers pictured in the rehabilitation propaganda literature were injured *behind* the front lines—not in valiant battles with the enemy, but rather in common, everyday military-industrial accidents. Davis, for example, was injured during combat training. Twelve of his ward mates experienced similar fates of being injured without ever seeing the battlefield.

Table 1 Twenty Disabled Soldier-Patients in Rehabilitation at Walter Reed Hospital, May 1917 to March 1918

Name	Rank	Age	Place of Birth	Occupation	Date of Injury	Incident	Diagnosis
W.D.	Cadet	21	Illinois, US	Student architect	(Pre-service 1914)	Polo injury/car accident	Injury, left knee
J.S.	Private	26	Poland	Plumber	05/18/17	Shot in left foot by drunken soldier	Pain, cannot bear weight
J.I.	Sergeant	35	Germany	Restaurant cook	07/13/17	Train accident	Broken, jaw and right leg
E. B.	Private	44	Kansas, US	Lawyer	09/26/17	Leg crushed by a truck, on duty	Fracture, left femur and tibia
L.J.	Corporal	20	Indiana, US	Butcher	10/08/17	Gun exploded during target practice	Amputation right leg
R.A.	Private	25	Tennessee, US	Unknown	10/15/17	Kicked by horse, at drill	Compound fracture, left tibia
H.M.	Private	20	New York, US	Unknown	10/28/17	Left elbow crushed by car	Partial resection, shoulder-wrist cast
P.S.	Private	30	Italy	Cabinetmaker	11/04/17	Foot crushed by US train	Amputation left foot
J.P.	Sergeant	24	Georgia, US	Lumberman	11/20/17	Explosive enemy shell, trench	Amputation left leg
C.G.	Private	26	Illinois, US	Manager, Swift and Co.	11/30/17	Enemy rifle bullet, trench	Amputation left leg
F.H.	Sergeant	36	Minnesota, US	Electrician	11/30/17	Shrapnel in arm/back, bullet to scalp	Loss of function, left forearm
I.E.	Private	23	Poland	Sewing machine operator	11/30/17	Machine gun bullet, trench	Fracture and paralysis, left humerus
E.B.	Private	26	New York, US	Architectural draftsman	12/10/17	Aerial bomb	Amputation right arm
J.G.	Private	24	Tennessee, US	Laborer	01/01/18	Rifle bullet to left hand	Adhesions and atrophy, left hand
M.B.	Private	22	Belgium	Factory worker	01/22/18	Grenade explosion, trench	Amputation left forearm
J.H.	Private	24	Missouri, US	Motor mechanic	02/25/18	Forearm crushed by US train	Amputation left forearm
P.C.	Sergeant	22	Ohio, US	Student	03/09/18	Foot crushed by truck	Amputation right foot
J.B.	Private	25	Russia	Locksmith, grinder	03/17/18	Grenade explosion, at drill	Amputation right arm
S.S.	Private	23	Mississippi, US	Unknown	12/28/17	Discharge of service rifle while cleaning	Amputation left arm
W.L.	Captain	22	Unknown	Laborer	01/22/18	Premature grenade explosion	Amputation left arm

Source: Table based on patient records found in the National Archives and Records Administration, Record Group 159, Office of the Inspector General, Walter Reed General Hospital, Box 1109.

Train accidents, friendly fire, faulty grenades, motorcycle accidents, and untamed horses caused a goodly portion of bodily damage—more, at times, than enemy combat. The high incidence of noncombat injuries can be partially explained by the military's heavy reliance on noncombatants, that is, soldiers who worked behind the front lines during the Great War. Industrialized war created an unprecedented demand for skilled laborers to work as engineers, railroad men, and truck drivers. The percentage of noncombatants to serve in the military went from 10 percent in the Civil War to over 60 percent during World War I.[22]

The paradox between the propagandized ideal that the true test of manhood occurred on the battlefield and the reality that most disabled soldiers never saw the front lines and instead sustained injuries from mundane, noncombat accidents precipitated a masculinity crisis. The crisis affected the soldiers who were made to feel ashamed by the occasion of their injuries as well as the commanding officers who worried that their patients might not be tough enough to endure the rehabilitation regime. Disabled noncombatants consistently felt the need to justify the honor of their wounds. The title of one noncombatant's memoir, "The Diary of a Dud," captured the depths of self-loathing in five short words. The army did very little to help disabled noncombatants cope with the nature of their wounds. Indeed, it refused to award medals of exceptional service such as the newly created Distinguished Service Cross and the Distinguished Service Medal to noncombatants. Such accolades were reserved only for men who exhibited extraordinary heroism on the battlefield.[23]

Given their lack of merit, disabled noncombatants became frequent targets of verbal abuse and their complaints were delegitimized simply by the fact that they had never seen battle. Sergeant Jerome E. Lane, recuperating at Letterman General Hospital, claimed that he and his fellow patients were "terrorized" by medical officers and frequently called "slackers" for their failure both as soldiers and as patients unable (or unwilling) to overcome their disabilities.[24] At Walter Reed, Private John S. Hall experienced similar verbal assaults. During his home side instruction with the 5th Engineers Training Regiment, Hall began to experience severe back and neck pains. He was sent to Walter Reed Hospital in March 1918 to recuperate. There he engaged in rehabilitative therapy, working in the greenhouse. Hall became restless during his weeks of treatment, unclear about the nature of his medical condition and unsure about how the greenhouse work would benefit him. "Nothing was told me," Hall told the OIG, "about my condition and I could not understand the medical terms [the doctors] used in examining me."[25] Compared with other soldier-patients at Walter Reed, Hall was well educated. He completed his junior year of high school before volunteering for service, a decision he believed would help him gain admission to West Point. Hall's lack of knowledge about his own medical condition, in other words, was not because he was uneducated; rather it was a result of a lack of communication on the part of his medical officers. Days after Hall complained in a letter to his parents about the poor care he was receiving—once the letter had been censored, intercepted, and read by his commanding medical officers—he received a dressing down from his treating physician.[26] The commanding physician accused Hall of being "yellow," that is, a coward, and threatened that "if he had the power, he would put [Hall] in the front line trenches in Europe." Hall's commanding officer suspected him of malingering, of avoiding being sent overseas by faking physical disability.[27]

Rehabilitation propaganda promoted such behavior on the part of its commanding officers by insisting that, regardless of whether the injury occurred while involved in combat or behind the front lines, physical disability emasculated men, making them incomplete.[28] Indeed the fear about the emasculating effects of disability ran so deep in the minds of rehabilitation officials that many of them overcompensated by arguing that disabled soldiers could achieve a higher form of manliness than able-bodied men.[29] "Only by the surmounting of obstacles," claimed Burgess, "does a man grow and attain his full mental and moral stature."[30] Dr. Harry Mock, an industrial surgeon and frequent *Carry On* contributor, wrote that America's rehabilitated disabled soldiers had become "better men than they were before."[31] Another *Carry On* author, a major in the US Army, argued that having a handicap gave a man "grit," allowing him to become stronger than overcivilized men who "lifted cream puffs and stuck pansies in their buttonholes." Still another article, titled "The Lucky Handicap," maintained "the man to feel sorry for is the poor fellow who is rich in everything but defects."[32]

The disabled man heralded in rehabilitation propaganda literature was the "supercrip," a man who, against all odds, surmounted his physical deficits and achieved spectacular career success.[33] Medical officials would frequently invite job-holding civilian disabled men to give bedside lectures to injured soldiers as a way of motivating them to engage in the military's rehabilitation program. Propaganda literature regularly featured men like Minnesota banker Michael J. Dowling, who lost both of his legs and his left arm from frostbite as a child. From there his life was one not of hardship, but of steady progress—he completed school, got married, had three children, launched a political career (he was speaker of the Minnesota House of Representatives from 1901 to 1903), and enjoyed leisure activities such as driving cars and riding horses. Dowling reportedly "laughed" at the word cripple, because he did not believe he was one.[34]

Certain soldier-patients, as well as their family and friends, challenged the ideals of heroism and manliness upon which the pursuit of rehabilitation rested. Maury Maverick, a lieutenant who suffered permanent injuries after fighting in some of the deadliest battles on the Western Front (e.g., St. Mihiel and the Argonne), concluded that "the line between coward and hero [was] . . . very indistinct." Maverick rejected what he called the "hero ideal," which he thought the "ruling classes" constructed in order to build armies, convincing men to take up arms and to ultimately "stop . . . [them from] thinking."[35] Maverick went on to live up to his name (the term "maverick" originated with his grandfather), earning a seat in the US House of Representatives in 1934 as a Texas Democrat who endorsed civil rights and antilynching legislation.[36]

Some of the most piercing criticisms, though, came from women who stood outside the male-dominated military establishment. One such woman was Miss Louise Irving Capen, a high school teacher from Rutherford, New Jersey, who learned of Private John S. Hall's case by word of mouth. In a letter to George Creel, director of the Committee on Public Information, the department responsible for nationwide wartime censorship of all published and unpublished materials, Capen complained that "it was not right—not even for the powers-that-be at Walter Reed—to down the enthusiastic patriotism of an American [soldier]." Instead of seeing rehabilitation as the key to making productive citizens, Capen argued that Walter Reed was making Hall into "an idler in a bathrobe." "He ought to be snatched from the idleness of the ward," she wrote Creel, "from the clutches of Walter Reed rule . . . and [taken to] a place where he could be of use." Contrary to what the propaganda wanted her to believe, Capen saw the military rehabilitation hospital as a place that promoted inactivity rather than productivity. In her mind Walter Reed had taken Hall, a heroic young man, and emasculated him. By "hammering away . . . at his character," she claimed, medical officials had turned him into an "indolent" with little interest in a career or his future.[37]

Women such as Capen became crucial to making patient grievances known. Because of the strict rule of censorship, whereby every letter that a soldier-patient wrote was read and vetted by medical officials, female hospital volunteers became key informants about what was happening behind the hospital walls. Indeed, most cases brought to the OIG concerning patient care on the home front came from publicly involved women who wrote letters to their congressmen and to army authorities.[38]

Rehabilitation officials found women like Capen to be "meddlesome." In a statement made to the OIG, Colonel Truby complained that "a lot of women come out here [to Walter Reed], offering their services and are doing more harm than good." "Most women," he continued, "come out here for curiosity, not to help." When one woman came to Walter Reed to volunteer her services, Truby reportedly told her that "if she wanted to help out to go down in the kitchen and wash dishes."[39] The main problem with these women, wrote Surgeon General Merritte Ireland, was that they engaged in too much "hero worship," an act that spoiled soldiers and intensified their feelings of dependency.[40]

In an attempt to keep society women out of his hospitals and ruining the reputation of the army Medical Department, Colonel Frank Billings, director of the OSG Division of Physical Reconstruction, sent a nationwide memo to more than a dozen local newspapers, urging "the residents of cities in proximity to the military hospitals [to] not injudiciously interfere with . . . the Commanding Officer." "The public should bear in mind," Billings asserted, "that the Commanding Officer and the medical personnel of the hospital have in mind . . . the best interests of the sick and convalescent men." At the same time, rehabilitation officials urged disabled soldiers to avoid fraternizing with such women.[41] In the illustration "When a Feller Needs a Friend," propagandists depicted society women as dangerous dimwits who kept honest men from finding employment. The disabled soldier, using a cane, makes a mad dash away from the women, avoiding the perils of female sympathy. If the "feller" was lucky, propagandists implied, then his male "friends" would keep him out of the clutches of such women—veritable Eves in the Garden of Eden.

Despite the rehabilitation officials' demonization of society women and female hospital volunteers, the message did not stick. While disabled soldiers undoubtedly heard lectures about the evils of charity and overly sympathetic women, they still showed inclinations toward motherly caregivers. Soldiers who wrote for the Walter Reed hospital magazine, *The Come-Back,* conveyed their desires for a "good old-fashioned" woman, often using humor as the preferred form of expression so as to avoid being censored. In one illustration, a bed-ridden soldier being cared for by two smiling nurses tells the reader in a dialogue balloon: "I don't wanna get well." Taking a further step toward acknowledging the fact that men in the camp liked having women take care of them, *The Come-Back* titled

When a Feller Needs a Friend

By Briggs

18 Illustration depicting how well-meaning society women could ruin the rehabilitation effort.
Carry On 1 (June 1918): 19.

the piece: "No Wonder Buddy Don't Care to Mend."[42] In another photo featuring three young nurses sitting on the hospital steps of Walter Reed, the *Come-Back* staff posed the question: "Wouldn't you be glad to be isolated here?" Admitting that it was pleasurable and desirable to be in close contact with young, available women, rehabilitating soldiers challenged both the tenets of medical professionalism (an outlook that prized objective interactions between healer and patient, especially ones between opposite sexes) and the goal envisioned by rehabilitation leaders.

Since an overwhelming majority of World War I soldiers were unmarried—draft boards frequently exempted married men from service—one of the most frequently expressed desires among disabled soldiers was finding a wife.[43] Contrary to the formulaic rationale provided by the framers of the WRIA that marriage would encourage men to become wage-earners and assist in bringing about social order after the war, rehabilitating soldiers at Walter Reed wished for companionship. Most of all, they wanted confirmation that they were still desirable despite their mutilated bodies. Certain disabled soldiers found the ritual of courtship easier than others. One group of amputee soldiers discovered that they could use their disability to their advantage when it came to "picking up girls." While taking the hospital car out for a ride around Takoma Park, the amputee soldiers found out that the "girls" who stood on the curbs waiting for public transportation often ignored able-bodied men, but "gratefully" accepted rides from disabled soldiers and happily piled into the hospital car.[44]

For those disabled soldiers who felt less assured, *The Come-Back* tried to assuage their comrades' fears about attracting the opposite sex. One article covering the Red Cross "Hop" used the event as a way to bolster the veterans' self-confidence, reporting that amputee men could easily socialize with women at dance parties. "Artificial legs and arms galore were in use," began the article. "Young ladies were quite impartial in their favors," the story continued "and if a man did not have the requisite number of arms and legs, they [the ladies] very agreeably 'sat out' that dance." By relaying a message of success and good cheer, *The Come-Back* staff sought to strengthen the confidence of men who were insecure in their courtship abilities. "Don't say it can't be done," the article stressed, "this . . . adage . . . applies more than ever to you disabled birds who think you're out of the game just because you've a leg or an arm gone."[45]

And yet, at other times, the disabled staff writers of *The Come-Back* displayed a more staid attitude about their prospects of getting "the girl" over able-bodied men. In one illustration a one-legged veteran with crutches holds a frayed, bundled-up piece of clothing in one hand and a letter in the other. The dialogue balloon above the amputee explains the scene: "I'll have to write her that it fits swell—Or it will break her lil' heart an' I ain't a goin' to do that. No Siree." With this cartoon, *The Come-Back* suggested that whereas an able-bodied man could demand more from a girlfriend or potential spouse—namely, that a knitted sweater would actually look like a sweater and not an unraveled hand muff—the disabled man needed to keep his expectations in check. Or as the headline of the illustration put it: "That Sweater from the ONLY Girl."[46] In other words, the disabled soldiers in charge of the *Come-Back* assumed that while most able-bodied men had many prospects for marriage, the disabled veteran should feel lucky if even one woman expressed interest in him and should thus reciprocate her signs of affection, even with false words of praise.

Racism and Rehabilitation

Despite the fact that African Americans made up 13 percent of the army (at a time when blacks made up only 10 percent of the population), they were gravely underrepresented both in rehabilitation work and its propaganda. Emmet J. Scott, the special assistant for Negro Affairs to the secretary of war, tried to get the OSG to hire black physicians, but in most cases the OSG refused.[47] In Jim Crow America, white physicians could treat black soldiers, but not the other way around.

Similar to the work itself, rehabilitation propaganda was overwhelmingly white in content and appearance. While rehabilitation officials devoted numerous articles to the Americanization of foreign-born disabled soldiers, they barely addressed the subject of disabled African-American soldiers. Only through the occasional photographic image would a reader of this literature even know of the disabled black man's existence. It was an existence without comment, for rarely if ever did rehabilitation officials actually provide stories of the African-American men who had been injured during the war.

Rehabilitation officials did not want to draw attention to the fact blacks and whites were not segregated in their hospitals, unlike virtually every other facet of military and civilian life. Worried that racial unrest could undermine the war effort, World War I army officials made black men serve in segregated units. Some African-American soldiers were farmed out to fight for France, wearing French uniforms.[48] But in certain circumstances, the financial demands and chaos of war made segregation impossible. Because rehabilitation was a new specialty, the Medical

19 When photographed, African-American disabled soldiers were pictured performing docile activities so as to convince readers that rehabilitation workers kept racial tensions and violence under control. *Carry On* 1 (August 1918): 24.

Department felt that it did not have the time or resources to construct and staff separate hospitals for blacks and whites.[49]

In order to quell the fears that racial violence would upend the rehabilitation process, medical officials in charge of propaganda consistently portrayed black soldier-patients as obedient, docile participants in their own recuperation. Photographers working for the Medical Department shot African Americans performing only the most restrained activities, such as basket weaving, light gardening, and reading. Taming African-American men had been a goal of the US Army since the beginning of the war. Even though black men had successfully taken up arms in the Civil War, the Spanish-American War, and along the Mexican border, the army of the First World War decided that in an era of Jim Crow and widespread race riots, armed black men, knowledgeable in the ways of combat and killing, would cause too much anxiety and potential unrest among white Americans. In order to keep rifles out of the hands of black men, army officials assigned most African-American soldiers to noncombatant units, making them perform menial, unskilled labor.[50] Using similar logic, rehabilitation officials discouraged African-American soldier-patients from engaging in athletic competitions and aggressive wrestling matches, reserving these activities for white men only.

But actual day-to-day relations between black and white soldiers proved volatile. Historian Jennifer Keene notes that throughout the war years, "individual brawls, street fighting, and outright rioting between white and black enlisted troops . . . were common."[51] At Camp Merritt, a port of embarkation in New Jersey, a race riot broke out on August 18, 1918, after a group of white soldiers who had just arrived from Mississippi insisted that the YMCA provide segregated facilities. A note was left for the YMCA secretary: "You YMCA men are paying entirely too much attention to the niggers, and white men are neglected. Because of this, if it is not corrected by sundown, we are coming to clean this place out. (Signed) Southern Volunteers."[52] By day's end, the once orderly camp fell into chaos, with white and black enlisted men assaulting one another. Three black soldiers were wounded. Another was shot dead. In retaliation, one black soldier pulled a pocketknife and slashed a white assailant's neck.

The same kind of racial violence occurred among US troops overseas. A participant in a hospital minstrel show, Private Guy R. Moore, Medical Corps, wrote of how, while entertaining a troop of black soldiers stationed outside of Angers, France, he nearly lost his life. "The negroes seem[ed] to like the clown stunts," Moore wrote in his diary, "but when the black-face parts took the stage we heard such expressions as 'Throw him out' and 'Shoot 'im.'" Moore and his fellow players cut the show short, slipping out the stage door and into an idling truck. While piling into the truck, the black soldiers "thronged to the scene." "They came toward us with a terrible fierceness," he recalled, "as if they meant to exterminate the gang."[53]

Similar racial tensions flared up in rehabilitation hospitals. In December 1917, a riot broke out at Letterman Hospital after a black man, Clifton Sobers, asked several white patients who were "monkeying around near his bunk" to leave. Upon hearing this interaction, two white hospital attendants dragged Sobers from his bunk and beat him until his "eye was cut open" and "a stream of blood could be seen from his bunk to the detention ward."[54] A black kitchen worker came to Sobers's defense and a hospitalwide brawl erupted, involving white guards and black soldier-patients. The army, an institution that readily appeased white opinion, transferred the two white attendants to a different hospital, while leaving the assaulted black soldier-patients at Letterman to continue their recovery.

Certain white disabled soldiers, intolerant of the fact that they had to share rehabilitation quarters with black soldiers, demanded transfers to all-white hospitals. In a letter to Mrs. B. R. Russell, a Walter Reed Red

Cross volunteer, Private James Cunningham wrote that he could no longer stand being in a ward with "fifty percent" of the men being "crazy Negroes." "Do you think," he wrote Mrs. Russell, "that this [desegregated ward] is a nice place for a man like me, the father of ten children . . . a patriotic volunteer for my country whose Father served in the Civil War and Great Grand Father served in the War of 1812?"[55] Involving the OIG and the OSG, Russell took it upon herself to see that Cunningham was transferred to an all-white hospital, claiming that Walter Reed's mixture of black and white patients was not the "right treatment" for a respectable white man such as Cunningham.[56]

Racism against African-American disabled soldiers did not end with the war. Although in theory the WRIA entitled black disabled soldiers to the same benefits as whites, the already entrenched racial discrimination that permeated America's health care system and labor markets made it virtually impossible for black soldiers to cash in on what was owed to them. In a letter to the NAACP, one black disabled veteran wrote that "since the war, some of the Southern crackers are using different means to keep we [sic] colored soldiers out of the hospitals and from getting vocational training."[57] At the war's end, the military began a rapid demobilization, including the dismantlement of the OSG's rehabilitation hospitals. The Public Health Service along with voluntary civilian hospitals contracted with the federal government to fill the void left by the OSG, offering its facilities for the rehabilitation of disabled soldiers. However, contrary to the Medical Department's insistence that disabled soldiers receive treatment without regard to color, the Public Health Service and the civilian hospitals involved did not feel obliged to follow the OSG's lead. Indeed, historian Vanessa Gamble notes that after the First World War, "many southern communities refused to hospitalize black veterans in the same facilities as white veterans."[58]

In the years to follow, integrated rehabilitative care gave way to segregation. By 1923, when the federal government instituted a system of veteran's hospitals (what we know today as Veterans Administration hospitals), the committee on hospitalization created one black veteran's hospital, the Tuskegee Veterans Hospital, while designating the remaining 26 hospitals for whites. Compared with the rate at which white veterans sought care, few black disabled veterans applied for benefits or signed up for rehabilitative care.[59] This trend was confirmed in 1935, when it was found that in eighteen states with the largest black populations, 91.5 percent of the rehabilitants were white.[60] While rehabilitation propaganda promised to enhance a disabled soldier's earning potential by providing training in skilled trades, African Americans could not surmount the systematic racism found in America's labor market. In the North, black men were largely limited to manual labor and low-grade work, such as janitors, barbers, and waiters. In the South, the demand for black agricultural workers was almost constant.[61] Not being able to break the racial barriers already in place prior to the WRIA, disabled black soldiers found the federal government's new program of rehabilitation was for white uplift, not their own.

Dependency and Defiance

Even though white disabled soldiers benefited more than blacks, rehabilitation still did not live up to its propagandized ideal. Despite receiving medical care and vocational training, many injured white soldiers found the postwar labor market unforgiving and intolerant of their disabilities. Historians Paul Dickson and Thomas B. Allen note that "countless veterans . . . resented the fact that civilian war workers had prospered in safety, their pay increasing by an average of 200–300 percent while soldiers . . . barely subsisted on military pay."[62] Sustaining a permanent disability made the discrepancy between the haves and have-nots even worse. In a capitalist marketplace, US industries were not expected to take responsibility for the needs of the disabled, but rather the disabled "were expected to adjust themselves to the dynamics of labor markets and to the occupational structure of an industrial economy."[63]

There were a few success stories. A North Carolina machinist who lost his leg in battle took classes in watch making and engraving during his rehabilitation. Once discharged from military service, he was offered a position in a jewelry store at $65 per week, $47 dollars more than his prewar weekly wages. In the end, the former machinist-turned-amputee jeweler opened his own business, making an even greater fortune. Rehabilitation officials commended him on his success and especially for mitigating "the need for continuous Government paternalism" and pay in the form of pensions.[64]

But for every success story, there were multiple accounts of failure. When George L. Cassiday returned home and tried to regain a railroad job he had held before being drafted to go overseas, he was turned down because of his war-related disability. He found an illicit means of making money, opening up a bootlegging business in the basement of the House of Representatives office building.[65] Other disabled soldiers returned to their former occupations but made less money. Consider the case of First Lieutenant Philip Nelson, injured in the fall of 1918 by "running into a Frenchman" while on a bicycle. After a full year (April 1919–March 1920) of rehabilitation for his fractured left leg, Nelson still suffered from joint stiffness, muscle atrophy, and a one-and-a-half inch "shortening" of the left leg, leaving him no other option but to walk, rather cumbersomely, with the assistance of a cane. Upon discharge from Walter Reed Hospital, he found work in civil engineering, his prewar occupation, but he could only perform desk work, for he was not able to go out into the field to conduct land surveys.[66]

Nonetheless, the officials in charge of rehabilitation propaganda aimed to convince its readers that a disability did not have to be a handicap. A frequent contributor to the propaganda effort, world-renowned bibliographer Douglas C. McMurtrie, assured disabled soldiers that their "handicap [would] not prove a serious disadvantage." As director of the Red Cross Institute for Crippled and Disabled Men in New York, one of the nation's first adult trade schools that assisted disabled men in finding employment by providing vocational training, McMurtrie believed that amputee veterans could perform so well in the workplace that even "a stranger would hardly know that [they were] crippled."[67] The medical staff assigned to the Division of Physical Reconstruction, he promised, would show injured soldiers how to "sink back into the mass of people as though nothing [had] happened."[68]

This denial of disability, from a magazine that devoted itself to the topic of disabled soldiers no less, was motivated in part by the concern that would-be employers would reject disabled soldiers outright. Rehabilitation officials had to convince businessmen, who were still adjusting to the new reality of workmen's compensation laws, that disabled men were no more at risk of work-related injuries than able-bodied workers. In an article written for *Carry On*, Samuel Harden Church, president of the Carnegie Institute, urged his fellow businessmen to stop expecting "physical perfection of one hundred per cent" from workers and to rethink the assumption that these disabled men were "defective members of [the] race."[69] In an era when social Darwinism held purchase on immigration and sterilization laws, Church and other rehabilitation proponents worried that disabled veterans might be mistaken for unsavory "degenerates."[70] British officials had similar concerns. In order to avoid confusion, the British army granted "eugenic stripes" to soldiers who were wounded in the line of duty. Soldiers were instructed to wear the stripes on their shirtsleeves so that the public could tell the difference between the worthy and the unworthy disabled.[71]

The rehabilitation effort was thus part of a much larger Progressive Era movement to stabilize the workplace and the nation's labor economy. Training rural migrants and immigrants in skilled labor, social engineers believed, would decrease employee turnover and strikes in the nation's urban centers, as well as improve workplace productivity. Early twentieth-century changes in the actual physical conditions of factory floors and mechanized modes of production created more sedentary jobs—work that required clerical, technical, and managerial skills, not brute strength. Rehabilitation officials saw the rise in corporate industry as a window of opportunity for the disabled, for physical deficits could be more easily overlooked in this kind of marketplace. To further the disabled soldiers' chances of gaining employment, rehabilitation officials insisted that men receive training in jobs that provided employment year round in "growth areas," that is, occupations that were so new they were not yet overcrowded. Of the couple thousand men who enrolled in vocational training, 50 percent received instruction in industrial trades, while others chose to take courses in clerical, agricultural, and retail work. A small percentage of disabled soldiers (16 percent) attended university to acquire professional degrees.[72]

Despite the fact that labor markets and occupational hierarchies were constantly in flux, rehabilitation officials assured disabled soldier-patients that by receiving training in skilled trades, they would be protected against such instability. Indeed, the propaganda literature attempted to convince its readers that the men injured from the war were fortunate to have been physically maimed, because only the disabled could partake in the vocational training that rehabilitation provided.[73] McMurtrie believed that disabled men could be financially better off because of their injuries, not despite them. "It would be cause for national pride," wrote McMurtrie, "if, in the future, [disabled] soldiers could date their economic success from the amputation of their limb lost in their country's service."[74] Rehabilitation officials refused to admit that the loss of a limb—or multiple limbs—really mattered. Instead, they followed the philosophy that "from the neck down a man is worth about $1.50/day; from the neck up, he may be worth $100,000/year."[75] As long as a man still had his intellectual faculties (and was white), he could supposedly make a fortune.

Rehabilitation officials made disability appear to be a temporary physical glitch, something from which men would fully recover with six months of hospitalization and rehabilitation. But disabled soldier-patients and their bodies repeatedly defied such lofty and unrealistic thinking. While rehabilitation officials attempted to deny the permanence and complexities of their injuries, disabled soldier-patients pushed back by lodging complaints to the OIG, extending their hospital stays, and trying to stage rehabilitation slow-downs.

Although many disabled soldier-patients found their hospital stays boring at best and terrifying at worst, a number of them wanted to extend their stays, complaining that the army had reneged on its promise to provide rehabilitation until each and every patient had been "maximally restored." This sort of complaint became increasingly common after the Armistice, when the army drastically scaled back its services, forcing medical discharges. Many disabled soldiers saw forced discharges as a way for the military—and the nation as a whole—to cut its loses. With the Civil War pension system defunct (except for those Union veterans who were still alive), medical rehabilitation and a parsimonious disability compensation system (a rate-adjusted scale determined by the War Risk Insurance Bureau) was all that injured soldiers of the First World War could rely on.

Demanding extended hospitalization was thus one way that World War I disabled soldiers fought for greater benefits. Private George Russo and Sergeant John B. Gordon were some of the first soldiers to engage in this kind of struggle. They made their grievances known to Congressman Royal C. Johnson (R-SD), a decorated lieutenant who absented himself from the House of Representatives to fight in the war. After meeting Russo and Gordon, Johnson penned a letter to the secretary of war in May 1920 insisting that the OIG investigate conditions at Walter Reed, stating that men "are asked to apply for discharges or receive discharges when they are unable to take care of themselves and no provision is made for their care by the War Risk Insurance Bureau." A six-month investigation ensued, resulting in over a dozen in-depth interviews with recuperating soldier-patients.

Russo, who had already spent twenty-two months recuperating at Walter Reed, complained that he was being "rush[ed] out of the hospital" before he was "fit to wear an artificial limb." Experiencing repeated blistering and pain, he found artificial limb wear nearly impossible. Using the logic that rehabilitation officials often espoused, Russo maintained that if he could not wear an artificial limb, he was not employable. "I have to earn my living when I get out," he told his medical officials, "[but] I cannot put any weight on my leg."[76]

Although already discharged, First Lieutenant Nelson contended that he still needed "hospital attention," but that because he had a wife and two children to support, he found it impossible to "spare the time" to receive care. "Unless some further treatment is given to my leg and hip," he told his interviewer from the OIG, "I shall be a cripple all the rest of my days."[77]

Disabled soldiers wanted to prolong their hospital stays because they worried that the monetary compensation provided by the War Risk Insurance Bureau was not enough to sustain them. One soldier-patient, Private Jeremiah Hurley, who at the time of his interview in the summer of 1920 had been recovering at Walter Reed since February 1919, told the OIG official that he still had "not reached the maximum treatment to be expected." Reporting that he would be wheelchair-bound for the remainder of his life (the records do not reveal the nature of his disability), Hurley maintained that he did "not desire to spend the rest of [his] life in a hospital," but that once he left Walter Reed, he did not "want to have to enter any other hospital for treatment." The main reason Hurley resisted discharge is that he believed that the army and the War Risk Insurance Bureau's medical board would give him a lower disability rating than he deserved. Although disability ratings were abstract algorithms that civilian insurers, military doctors, and the War Risk Insurance Bureau used to assign monetary worth to body parts lost in war, these numbers made very real and concrete differences in the actual lives of disabled veterans. The distinction between a low and high disability rating could mean the difference between receiving $20 per month and $120 per month. Hurley resisted being discharged because he believed that he deserved "double total disability," a rating that would have given him a $200 per month allowance. "What I want is justice in this disability," he told one OIG investigator. "I see a man with loss of one leg or a foot receive total disability," he observed, "[but] I need a constant attendant after discharge [and] the allowance of twenty dollars per month is not . . . sufficient."[78]

Medical officials accused men like Hurley of "gold bricking," bilking the US army and the federal government of money. As the Civil War pension roll grew, rehabilitation officials worried that the country could not afford to provide similarly generous benefits to the millions of newly disabled soldiers. To discourage World War I soldiers from feeling that they too deserved pensions, rehabilitation propagandists portrayed soldiers who felt entitled to government "hand-outs" as "parasites," men who sucked the democratic spirit out of American society.[79] The ever present anxiety in the minds of rehabilitation officials was that World War I disabled soldiers would wish for (and perhaps demand) a life of leisure.

In reality, though, most of the men who found themselves disabled and rehabilitating at Walter Reed and Letterman came from working-class backgrounds (for a sample of occupational backgrounds, see table 1). The economics of everyday life necessitated that many patient-soldiers find employment, whether disabled or not.[80] In addition, a life of hard work—whether skilled or unskilled—was what most of the men who came through the Medical Department for rehabilitation services were accustomed to. To take one example, Private Lee F. Steinbacker, admitted to Letterman General Hospital at the age of twenty-five on August 8, 1917, left grammar school in the eighth grade in order to begin work on a farm near his home in Kansas, helping to support his parents and five siblings. After farming from the age of 14 to 20, he worked as a "general laborer" and eventually became a "nurse in a civil hospital" before enlisting in the Ambulance Corps in the spring of 1917.[81]

The problem posed by disabled soldiers who wanted to remain in the hospital for extended periods of time was that they seemed to be following in the footsteps of their Civil War predecessors, men who took up permanent residence in the country's National Homes. The negative portrayal of the National Homes for Disabled Veterans can best be seen in an ink drawing featured in *Carry On*. The central figure of the illustration, a disabled World War I soldier, stands in a trench, lined with money bags, sacks of federal dollars tagged for charity to disabled soldiers. The assumption here is that the federal government had dug itself into a financial hole after the Civil War by giving injured veterans generous pensions without expecting them ever to go back to work.

The illustrator, H. T. Webster, made this point even clearer by situating the "soldier's home dugout," a representation of the Civil War era National Home, in the belly of the trench, the place where only cowards retreated, seeking protection. Webster's intended linkage between the "old" Civil War pensioner and weakness is also confirmed by his use of the poison cloud on the horizon that reads "worn-out notion about the cripple." Contemporary viewers would have associated poison clouds with "feminine" forms of fighting, a battle strategy used by the Germans as a way to avoid hand-to-hand combat with enemy forces.[82] In other words, disabled World War I soldiers were supposed to reject all weak and feminine forms of support—pensions, extended hospitalizations, governmental support—and instead "go over the top," following the instructions of the signpost situated in the focal point of this illustration: "to a man's land: a job."[83]

To further persuade disabled World War I soldiers that the WRIA served their best interests and that it was an advancement rather than devolution in veteran welfare policy, rehabilitation officials encouraged Civil War veterans to provide antipension testimonials. Colonel William Thomson, a Union veteran who had lost his arm and had spent time in a National Home in Virginia, informed the new generation of disabled

They Don't Want Your Charity—They Demand Their Chance

By H. T. Webster

20 An Illustration making the argument that rehabilitation provided better treatment to veterans than the Civil War pension system. *Carry On* 1 (August 1918): 4.

soldiers that "one of the saddest mistakes made by the Government after the Civil War was the . . . soldiers' homes [where] disabled men . . . if they desired, [led] an idle life while enjoying the pension to which they were entitled."[84] Colonel Thomson's description of the Civil War system, however, could only be seen as a criticism if idleness and long-term care were understood to be vices rather than benefits.

Certain disabled World War I soldiers interviewed for the Russo-Gordon case applauded Walter Reed and its rehabilitation regime. But their words of tribute still bore witness to the fact that, contrary to the propaganda, there was no quick fix to physical disability. First Lieutenant Harold H. Tittman, who was discharged from Walter Reed in May 1920, wrote later that summer that "he [could] only speak of [his] seventeen months at Walter Reed Hospital in words of praise." Although Tittman had not requested his own discharge, "it was suggested" to him. "At first I was somewhat surprised," he wrote, "that I should be signalled [sic] out as cured when there seemed to be so many around me who were in so much better condition and who were not up for discharge." Tittman concluded that he was probably chosen because he had "lost interest in physiotherapy, [his] attendance there being more than desultory." Although convinced that his discharge was for his own good, he told the OIG that his right fingers were still ankylosed and that the shrapnel wound in his left leg had recently reopened and begun draining. "I expect to be picking dead bone out of my leg for some years to come," he calmly wrote, "all I [can] do is to keep the wound clean and bandaged."[85]

For all of their attempted protest and resistance, however, many disabled soldiers found themselves discharged against their will and with disability ratings lower than they felt they deserved. With the Medical Department under intense pressure to demobilize, by 1920 military medical officers began constructing rationales for why extended hospital stays were deleterious to their patients' health. They even went so far as to describe the desire to remain in the hospital as a disease called "hospitalism." Lieutenant Colonel William L. Keller, in charge of the disposition board for examination for discharge at Walter Reed told the OIG that he had to discharge many patients because "the depressing atmosphere of a hospital [was] injurious to [their] health." Military physicians used the diagnosis of hospitalism, much like malingering, to avoid taking responsibility for medical failure, to explain away all those patients who did not recover or who did not engage in rehabilitation and readily adhere to authority. Hospitalism placed the blame on the patient, making him responsible for not responding to treatment quickly enough, for not being compliant, and for not knowing that, after a time, he had to stop asking for further assistance.

*   *   *

On a train bound for a field hospital in western France, Lieutenant Maverick lay on a cot next to a German soldier who had been wounded in the

Meuse-Argonne Offensive. The German soldier had been gravely injured. "Most of his face had been shot off," Maverick described. "He would sip a little milk," Maverick recounted, "but the blood would trickle down into his stomach and he would puke blood and milk." Maverick remembered his bunkmate not simply because of the horror of his injuries, but because he told Maverick about the Bismarckian system of pensions and social insurance granted to every veteran of war. "If I do not die," the faceless soldier told Maverick, "I will take a long vacation [at the government's expense] and get well."[86] The soldier died within an hour of saying this, but his story of veteran benefits lived on in Maverick's mind. As soon as he returned to the United States, Maverick became a leading figure in the fight for improved veterans benefits for World War I disabled soldiers—a fight that would ultimately lead to the violent 1932 Bonus March on Washington, where aggrieved veterans of the Great War demonstrated on the Capitol grounds demanding just compensation for their wartime service.[87]

The protests that culminated in 1932 grew out of soldier complaints that had begun during the First World War. Although the propaganda produced by rehabilitation officials attempted to convince wounded soldiers that Bismarckian benefits and Civil War pensions would be injurious to their health and postwar livelihood, soldiers wounded from the war sensed injustice from the outset of their recovery. As a system of medical welfarism predicated on the belief that disabilities could be cured—and disappear—in a matter of months, rehabilitation failed to live up to its billing. Feeling short-changed, white and black disabled soldiers protested against the system, for, as they rightly suspected, Wilson, Congress, and the Progressive reformers behind the WRIA legislation supported rehabilitation as a means to make a clean break from the past as well as a future of continual veteran demands.

SEVEN

Rehabilitating the Industrial Army

Although the WRIA and the rehabilitation regime began as an experiment—conceived of and passed by Congress during a time of war—it persisted after the Armistice, informing peacetime policy for America's disabled citizens. Indeed, a little over a year after the war ended, Congress passed the Civilian Vocational Rehabilitation Act (CVRA) mandating that federal monies be made available to victims of industrial accidents, helping them to recuperate and find new employment. Congress reauthorized the CVRA every four years until it became permanent legislation in 1939 as a part of Franklin D. Roosevelt's New Deal.

To Julian Mack and the other framers of wartime rehabilitation policy, the CVRA was a monumental achievement, for they rightly saw it as an expansion of the WRIA to the civilian sector, a step toward making worker protection a federal matter. Prewar social reformers fought hard to federalize worker protection laws, but most legislation such as workmen's compensation remained on the state level. Not until rehabilitation was wrapped in the cloth of wartime patriotism—a program billed as necessary for the welfare of disabled soldiers—did it receive overwhelming congressional support. In short, legislation that passed during the Great War for the disabled soldier was a crucial stepping-stone in making rehabilitation available to the disabled civilian laborer from 1920 onward.

But in the process of passing the CVRA, medicine became decoupled from vocational training. While the CVRA paid for vocational counselors to make home and worksite visits, providing occupational training and job placement services to injured workers, it did not make health care services available to the industrially disabled.[1] Although the reasons for withholding health care from the CVRA legislation were complex—from fears that the United States would look like a socialist state to worries about cost—the end result was clear: the United States would become a nation that provided both medical and vocational rehabilitation to its disabled soldiers (a system that continued after the First World War through the Veterans Administration) while granting only the latter to its nonmilitary disabled citizens. America's two-tier system of socialized medicine for soldiers and privatized health care insurance for the rest of its citizens persisted for the remainder of the twentieth century.

Contrary to the wider trend throughout the twentieth century of medical doctors opposing federalized health insurance, World War I rehabilitation physicians fought against the passage of the CVRA because they believed the bill should have included health care for industrial accident victims. Orthopedic surgeon David Silver put it the following way: "it seems important that we should . . . pay attention to the industrial [soldiers], for the problem of military surgery and that of industrial surgery are essentially the same."[2] From the outset of the war, the OSG intended to maintain full control over rehabilitation, not ceding any power to vocational educators who were making important inroads into the federal government at the time. The Federal Board of Vocational Education (FBVE), for instance, had been established in February 1917, just before the war. Soon after he created the army's Division of Physical Reconstruction, Surgeon General Gorgas sent the secretary of war a lengthy plan for rehabilitation stating that the military would take "civilian patients and rehabilitate them" alongside disabled soldiers.[3] Most significantly, the OSG took charge of drafting the first version of the Federal Vocational Rehabilitation Act, the predecessor to the CVRA.[4]

Ultimately, the OSG and its rehabilitation medical officers failed to broaden their sphere of influence, leaving their work marginalized to the military. Many factors played a role in the OSG's lack of success. First and foremost, nonmilitary policy makers worried that the line between civilian life and the armed forces would be blurred, making the United States into a military state. The OSG never provided a clear plan for how civilians would remain civilians while under the military's care. Equally important was the fact that in World War I America, health insurance had become widely unpopular. Policy historians blame the failure of early twentieth-century health care initiatives not only on commercial insurance companies and the mainstream labor movement, but also—and most important—on the American Medical Association (AMA).[5] In 1919, the AMA officially declared its opposition to government health insurance and so-called state medicine, characterizing all forms of compulsory health insurance as a Teutonic legacy of Otto von Bismarck, a depiction fueled by George Creel's wartime propaganda initiative.[6] The story of the OSG's attempt to extend rehabilitative care to the civilian sector—a movement spearheaded by civilian physicians who served as volunteers for the military medical corps—complicates the narrative that the medical profession was uniformly opposed to all forms of health care insurance.

The CVRA passed with ease precisely because it did not include health insurance. As a program controlled and operated by educators and social scientists, vocational rehabilitation under the CVRA appeared to be the antithesis of welfare, purportedly making individuals independent of the state, not dependent on it.[7] The OSG could not convince Congress that the Medical Department's rehabilitation program was not a form of high-cost, health care welfarism. Postwar reports coming from Walter Reed and Letterman Hospitals estimated that the average hospital stay for disabled veterans exceeded one year. By war's end, the army's rehabilitation program—a regime that was supposed to replace the Civil War system of pensions and long-term care facilities, such as the National Soldier's Homes—looked more like its predecessor than a solution to it. Thus, no matter how hard the medical officers in charge of the army's rehabilitation program fought to extend their program to the industrial workplace, Congress did not want to commit to the expense. The CVRA was a much cheaper solution, for it did not require institutionalization, hospitalization, or a long-term commitment to the health of the nation's laborers.

The Prewar Origins and Popularity of Vocational Education

In many ways, 1910–20 was a decade devoted to the uplift of the white male industrial worker. Beginning in 1911, states across the nation instituted workers' compensation laws as a way to improve industrial conditions. Lagging behind Great Britain, Germany, and France, all of which had implemented a no-fault workmen's compensation system by the end of the nineteenth century, the United States had the worldwide reputation of "maiming, mangling, and killing those who attempted to earn their bread."[8] With workers' compensation, each state required that private employers meet minimum industrial safety standards. In addition, when an employee was injured, business owners were obligated to pay a portion of the worker's preinjury wages and cover a visit or two to the doctor's office.[9] Because a wide array of political interest groups, from the American Federation of Labor to academics and big business, supported some system of industrial accident reparations, every highly industrialized state in the country had passed its own version of a workers' compensation act by 1915.[10]

Yet US workers still lacked the kind of safety net that their European counterparts enjoyed. In 1911, Chancellor of the Exchequer David Lloyd George persuaded the British parliament to pass the National Insurance Act, providing sickness insurance funds for most laborers between the ages of sixteen and seventy. Lloyd's legislation was based on Germany's sickness insurance plan that had been in operation since 1883, when Chancellor Otto von Bismarck instituted sweeping reforms—from old-age pensions to accident insurance—in order to discourage political radicalism among German laborers. By 1911, Bismarck's health plan covered 77 percent of all German workers.[11]

Inspired by Britain's National Insurance Act, a group of Progressive Era academic economists from the American Association for Labor Legislation (AALL) drew up a plan that would provide US industrial workers with federally funded long-term medical care, sick pay, and a small death benefit. In his book titled *Social Insurance* (1913), AALL member Isaac Rubinow claimed that the United States was twenty-five years behind Europe in insuring workers against illness.[12] Come World War I, Rubinow's arguments urging America to follow a German precedent created a backlash against the AALL. Historian Beatrix Hoffman writes that during the war "health insurance advocates faced the delicate task of pushing a German model while downplaying its origins."[13] The balancing act became too difficult to maintain. In the height of America's postwar red scare, the AALL's 1919 attempt to get health care legislation passed in the state of New York ended disastrously. Opposing politicians claimed that the bill was part of a socialist plot, and they killed it before it even reached a vote on the House floor. Defeated, the AALL ceased to promote health care legislation from that point forward.[14]

Vocational education (and eventually rehabilitation) turned out to be a much less contentious—and less beneficial, at least from the worker's perspective—form of worker relief than health insurance. Advocates argued that vocational training in secondary schools could solve many of the nation's social and economic ills. These proponents believed that vocational schooling "would integrate immigrants into the labor force, slash worker turnover, lessen labor conflict and social alienation, reduce unemployment and increase occupational opportunities for poor and working-class youth."[15] Reformers, labor leaders, social workers, farmers, and businessmen of various political stripes rallied around the cause. In the late nineteenth century, the National Association of Manufacturers (NAM) became a leading advocate of institutionalizing vocational training. Although many Americans viewed the Bismarckian system of sickness insurance with disdain, they stood in awe of Germany's economic success in the international marketplace. The NAM attributed Germany's achievement to its system of vocational schools, institutions that taught its students finely calibrated skills to meet the needs of specialized commerce and industry. In order to keep up with their international competitors, several large US manufacturers, including General Electric, Westinghouse, and US Steel, created onsite "corporation schools" to train workers in machine repair and low-level management, as well as skills in sales, clerking, and typing.[16]

Not wanting to bear the burden and cost of maintaining such schools, the NAM joined forces with American educators in the hopes of making vocational education a federally funded program. Not all public educators and welfare reformers were sympathetic to the NAM. Some vocational education advocates, such as John Dewey, believed that trade education should be integrated into the public school system so that all American school children could become "masters of their industrial fate."[17] A champion of democratic education, Dewey feared that, left in the hands of the NAM, vocational education would perpetuate the existing industrial order, channeling lower-class students into predetermined occupational slots with no opportunity to move up the vocational ladder.[18] Others of the more romantic persuasion wished to use vocational training as a way to restore the "creative impulse in industry" so that America's profit economy might be overthrown entirely.[19]

The NAM found a champion for its cause in vocational educator Charles R. Prosser, director of the National Society for the Promotion of Industrial Education (NSPIE).[20] The NSPIE was formed in 1906, when the Douglas Commission, named after William L. Douglas, the governor of Massachusetts, published the results of a study that found that 25,000 of the state's adolescents did not attend school for reasons of boredom and dissatisfaction. Of these 25,000 teenagers, 33 percent went to work as unskilled laborers, and the remaining 67 percent stayed at home, "idle." The NSPIE was charged with the task of establishing vocational schools in the state in order to help children develop skills for industrial society. The NSPIE believed that parents would force their children to stay in school until the age of 16, if such a practical form of education existed.[21]

Unlike Dewey, Prosser advocated the establishment of separate training schools, detached from the public school system. Prosser believed
that through the new sciences of educational psychology and sociology,
vocational educators could match the traits of each individual worker
to the most suitable job. Indeed, to him, the problems between bosses
and workers were due to "individual maladjustments," not class conflict.
To remedy worker unrest, Prosser and the NSPIE set out to educate and
"adjust" young workers, helping them find their "appropriate places in
the division of labor."[22] In other words, there was an ideal job for each
individual worker, and it was the vocational educators' task to find that
perfect fit.

Having enjoyed success in his home state of Massachusetts, Prosser,
with the support of the NAM, set out to secure federal support for universal vocational training. In 1911, he helped Senator Carroll S. Page (R-VT)
put forth a vocational training bill that called for an annual appropriation
of $5 million for vocational instruction in the public secondary schools,
$4 million for support of state-controlled agricultural high schools,
$1 million for agricultural experiment stations at state-controlled schools,
and additional sums for land-grant colleges.[23] Because of inherent urban-
rural tensions, the bill failed: vocational education was viewed as urban
aid, while agriculture funds helped rural populations.

Not until the NSPIE brokered a deal with Senator Hoke Smith (D-GA),
a long-time advocate of agricultural extension work, did vocational education win congressional support and come to enjoy a permanent place
in US educational and worker welfare policy. As a progressive, Senator
Smith thought that both agricultural extension work and vocational
education could facilitate his state's economic growth.[24] Ultimately, he
hoped that vocational training would stem the northward migration of
black laborers from the South. Following the "cast down your bucket
where you are" philosophy espoused by Booker T. Washington and the
Tuskegee Institute decades earlier, Smith believed that if southern blacks
received training, they would be content to remain in the region of their
birth, thus assuring the South of an improved labor supply.[25] The Smith-
Lever Bill passed Congress in 1914, leading to the creation of the Commission of National Aid to Vocational Education.[26]

The status and visibility of vocational education grew as America's
engagement in the war with Europe became more imminent. In December 1916, President Wilson urged Congress to provide increased financial support to vocational education, arguing that it was a necessary
means of war preparedness. European armies already involved in the war
were asking their soldiers to perform tasks normally reserved for civilian

contractors and laborers.[27] The Great War was a total war, with belligerent nations utilizing military, agricultural, technological, industrial, and human resources—all geared toward the aim of keeping the war machine going until eventual victory. If the United States entered the war, Wilson reasoned, vocational education would help the nation in its efforts to increase military industrial productivity. Congress heeded Wilson's advice and in February 1917, both houses passed the Smith-Hughs Bill, appropriating funds for the creation of the Federal Board of Vocational Education (FBVE), an agency that consisted of the secretaries of agriculture, commerce, and labor, as well as the commissioner of education, and three members appointed by the president.[28]

The FBVE named Prosser as its executive director. The law's language called for a federal-state partnership, whereby the executive director allocated federal funds to state vocational education schools that trained adolescents in agriculture, trade, industry, and home economics.[29] As soon as the United States declared war, the FBVE made it mandatory for state school authorities to train conscripted men in manual trades before they were drafted into the war. Prosser estimated that the US Army would need two hundred thousand highly trained mechanics to fight the war. By war's end, the FBVE trained over 62,161 conscripts in war production jobs such as automobile maintenance, engine repair, and radio operation.[30]

Medicine vs. Education

When the WRIA passed in October 1917, the clause that granted every disabled soldier rehabilitation, reeducation, and vocational training remained vague, allowing for various interpretations. The ambiguity of the WRIA set the stage for a vociferous debate between vocational educators and medical doctors, with each claiming professional expertise over the rehabilitation process, especially the endpoint of reemployment. Vocational educators believed that medicine should be decoupled from the latter stages of occupational guidance, whereas medical doctors contended that, for the sake of continuity of care, a physician should oversee all phases of rehabilitation.

From the beginning of the war—even before the WRIA had been passed—the Army Medical Department understood rehabilitation to be a medical process, one that began with surgery and ended with job training in curative workshops. The medical officials in charge of wartime reha-

bilitation, namely, orthopedic surgeons, had been incorporating vocational training into their practice since the 1880s, when the first so-called
hospital schools were opened in Boston, New York, and Philadelphia.

The OSG's conviction that rehabilitation belonged in the hands of
medical men was bolstered by examples from abroad. From the outset
of America's engagement in the First World War, the OSG sent physicians oversees to apprentice with European doctors engaged in the aftercare of maimed soldiers. While abroad, physician representatives of the
OSG attended the annual Interallied Conference on the Rehabilitation
of War Cripples in Paris in May 1917. The conference included representatives from all over the globe, including Australia and India. Paper
topics ranged from medical gymnastics and prosthetics to the need to get
wounded agriculturalists back to work for the sake of economic survival.
Trumpeting the superiority of the medical profession, members of the
Paris conference concluded that the "reeducation of the wounded" was a
two-step program of "functional restoration" followed by "occupational
reeducation," both phases of which "should be placed under the chief
direction of . . . doctors" who would be advised by a team of technical
workers, sociologists, and educators.[31]

Surgeon General Gorgas took to heart the multidisciplinary thrust of
the Paris conference and appointed an interdisciplinary team of medical and nonmedical experts—orthopedic surgeons, industrial physicians,
civilian tradesmen, and social workers—to establish the newly created
Division of Physical Reconstruction (DPR) in August 1917.[32] The OSG
planned to recruit men "from commerce, industry, and agriculture;
crippled men from civil life; capable injured men returned from active
service who have recovered; and successful teachers and mechanics selected from the draft army." In this setting, Gorgas maintained, education *and* medicine would work "hand in hand" for the benefit of the
patient.[33]

The surgeon general made known to the public his intention to control all aspects of rehabilitation, outlining the nature and structure of the
DPR in the *New York Times*. According to Gorgas, rehabilitation included
not only "physical rehabilitation" (i.e., the application of all medical
and surgical measures to restore the disabled to as near normal as possible), but also "vocational training, economic, and social supervision,
and replacement in industry . . . to [make] the disabled man a productive
agent."[34] By November of that year, Gorgas sent a memo to the secretary
of war, Newton D. Baker, telling him that in addition to Walter Reed and
Letterman, he planned to build sixteen military rehabilitation hospitals

that would be evenly distributed throughout the United States and to retain some of these facilities "for the training of [civilian] men handicapped by industry."[35]

From the OSG's perspective, the army's rehabilitation system was where the recuperation of disabled industrial workers should take place. Most early twentieth-century civilian hospitals rejected patients who required long-term care. "No general hospital worth its salt," writes historian Rosemary Stevens, "wanted to treat the . . . [chronic] conditions presented by a majority of veterans."[36] Civilian hospitals wished to attract middle-class, private obstetrical and pediatric patients, not working-class disabled veterans. Moreover, most civilian hospitals lacked the technology, staff, and structural support to deliver rehabilitative care for adults. The army's rehabilitation facilities promised to be larger in scale and production than any civilian hospital in existence. Military architects drew up plans for thousand-bed hospitals with each facility offering physical therapy, occupational therapy, and curative workshops.[37]

Prominent civilian physicians shared the belief that the army offered the best place for rehabilitation care as well. Many of the highest ranking medical officers involved in the DPR were leaders in academic medicine; few, in other words, came to the job as career military surgeons. Before becoming director of the DPR at the age of sixty-three, Dr. Frank Billings served as president of the AMA (1902–3), held the position of dean of the Department of Medicine at Rush Medical College, and was heading up an advisory committee to build a medical school on the south side of Chicago, later to become the University of Chicago Medical School.[38] Seeing his work in the army as an extension of his ambitions at Rush Medical College, an institution that later became well known for industrial medicine and surgery, Billings believed that medical care was essential to "conserving industrial manhood" and that military hospitals were leading the way in accomplishing this goal. Major John Shiels, a physician who worked at Letterman General Hospital during the war, maintained that patients received better care there because there was a "collaboration and coordination" not seen in civilian facilities. "If a man comes here with an infection," Shiels pointed out, "he has at his disposal a general practitioner, specialist and a surgeon, whilst outside in civilian life, the bringing together of this type takes a period of time."[39]

As the physicians of the DPR focused their efforts on establishing a network of rehabilitation programs for the army, they frequently pointed out that civilian laborers were more at risk of becoming disabled than military soldiers.[40] "There is another soldier," Dr. Harry Mock, a lieutenant colonel, maintained, "the industrial soldier . . . who becomes disabled and wounded without the glorification that comes from such wounds when received on the battlefield."[41]

In 1917, the Department of Labor estimated that there had been 875,000 nonfatal accidents that year, leaving approximately 74,530 workers permanently disabled.[42] By contrast, the War Department reported that out of the 1,046,533 men enlisted in the Armed Services, only 24,541 soldiers had been discharged for disability during the first year of military engagement.[43] As a physician working for the DPR, Major P. B. Magnuson argued that industrial medicine "should do away with poor medical care" and that military rehabilitation hospitals offered superior services.[44]

Just as social insurance reformers and vocational educators utilized the war emergency to push federal legislation through Congress, so too did industrial physicians and orthopedic surgeons attempt to secure federal monies (albeit through the war department) for the expansion of their own specialties, providing government-funded medical care to industrial accident victims. Compared with their European counterparts who could find work in government-run accident hospitals and rehabilitation institutions, US physicians and surgeons interested in industrial medicine had to rely on private paying patients (a strain for the working class) or find employment with industrial corporations themselves.[45] Some physicians worked for the Baltimore and Ohio Railroad; others found employment with the US Steel Corporation.[46] Although the systems of health care and accident insurance in Germany and Great Britain were complex (and, at the time of war, separate from military rehabilitation), both countries provided a more comprehensive plan of medical coverage for the working class than the United States.

Before sending a bill to Congress that would grant the OSG both full control over the rehabilitation of disabled soldiers and permission to begin providing care for civilian disabled workers, Billings held a weeklong conference, inviting all government and nongovernment organizations, departments, and professionals interested in rehabilitation. The list of participants included the surgeon generals from the navy (William C. Braisted) and US Public Health Service (Rupert Blue), representatives from the Medical Section of the Council of National Defense, the Employees' Compensation Bureau, the FBVE, the Department of Labor, the Treasury Department, the American Federation of Labor, the National Manufacturers Association, and the Red Cross Institute for Crippled and Disabled Men (headed by Douglas C. McMurtrie).[47] The OSG's tentative bill urged Congress to create a rehabilitation advisory board giving the army Medical Department executive control over all other representatives of the rehabilitation movement. In order to pave the way for a "permanent plan for the rehabilitation of all civilian disabled workers in the future," the OSG intended to provide vocational rehabilitation without cost to the Employees' Compensation Commission (ECC), a federal department created in 1916 to cover and adjudicate current and retroactive injury claims filed by government employees.[48] The ECC covered letter carriers employed by the US Post Office, laborers who had completed the construction of the Panama Canal, and the staff of the Alaskan Engineering Commission, who were laying rail to connect the continental United States to the resource-rich Territory of Alaska.[49]

The conference ended with more disputes than resolutions. Not wanting medical physicians to take control of vocational training, Smith and Prosser of the FBVE left the meeting ready for a fight. Using his prerogative as the congressional chairman of the Joint Committee on Education and Labor, Smith submitted an FBVE-sponsored rehabilitation bill to Congress and arranged for hearings that featured the testimonies of Prosser, T. B. Kidner, the vocational secretary of the Invalided Soldiers' Commission of Canada, and McMurtrie. The FBVE bill looked very similar to the one drafted by the OSG except for two crucial differences. First, the FBVE put themselves in charge of vocational rehabilitation, wresting it from the hands of the OSG. Second, and most crucially, the FBVE removed the clause that demanded coverage for disabled civilian workers.

While several congressmen expressed the need to include civilian disabled workers in the proposed rehabilitation legislation, the primary framers of the FBVE bill—Senator Smith, Representative William Joseph Sears (D-FL), and Representative Simeon Davison Fess (R-OH)—worried that inclusion of the industrial worker would jeopardize speedy passage of the bill. Sears maintained that, while he supported the cause of the industrial worker, he wanted to see legislation that covered the injured soldier first. "After we show the country the good that can be accomplished" in rehabilitating the disabled soldier, Sears reasoned, "the civilian bill will follow." Dr. R. M. Little, of the ECC, pushed Representative Fess on the issue and pointed out that while the FBVE was "doing something for all the normal young people of the country," it left the "the disabled . . . at the bottom, untouched."[50] Fess agreed, but, speaking directly to Little, explained his rationale the following way: "I advised against including [the amendment for the industrially injured] at this time, upon the conviction that under a great patriotic national impulse . . . there will be scarcely any opposition to this bill . . . but I do fear, if we should include the thing you want, which I am very much in sympathy with, we may have some trouble."[51] Fess feared a political backlash similar to that which the AALL was experiencing. Fess and his colleagues worried that if they proposed a federally funded welfare program for disabled workers, they would be accused of being agents of a pro-German socialist state, a charge that could end even the most promising political career.

Instead of pushing through legislation that would cover the civilian industrial disabled, framers of the FBVE-sponsored bill made it their primary goal to delegitimize the medical profession's involvement in vocational training. Vocational educators did not wish to exclude medicine from the rehabilitation process, but they did want to marginalize it, making it the first and not the final step toward full vocational recuperation. "If it is attempted to put this work into the hands of army doctors," McMurtrie told Congress, rehabilitation "will fail."[52] In his mind, medical doctors could not address the "social and economic rehabilitation of crippled men." Prosser was of a similar mind, insisting that a clear delineation between medicine and vocational training be made. Unlike medical authorities who were influenced by the Paris conference and who divided rehabilitation into two medically based phases (functional and occupational rehabilitation), Prosser proposed a more complicated arrangement. First the soldier was to engage in "bedside occupation"—a phase necessary to "buck up" the soldiers—which would be followed by training in the curative workshops housed at military hospitals. Finally, after the soldier had been discharged from medical care, he would, according to Prosser, "begin the more *serious* job of vocational reeducation" (emphasis added).[53]

Vocational educators and their political proponents also maintained that long-term hospitalization would be deleterious to the soldier's health. Smith and his allies contended that the OSG's maximalist approach to rehabilitation—a plan that kept patients under medical care until they were "fully recuperated," medically, socially, and vocationally—would at best be detrimental to a soldier's state of mind and at worst unachievable. "Experience has demonstrated," Prosser maintained, "that disabled men while under hospital treatment . . . fall into a state of chronic dependence, characterized by a loss of ambition."[54] Vocational education advocates remained skeptical about how much medicine, and the hospital, could benefit the soldier who needed to be reintegrated into his home community and society. "When the physical man is cured," Smith argued, "the work of the surgeon ends." Smith likened hospitals to machine shops, places where bodies were materially restored. Consistent with this view, he believed that "little practical vocational work [could] be accomplished in the hospitals," and that "the real work [began] after the patient [left] the hospital." At his angriest, Smith asked his fellow congressmen, "What training [does] the surgeon have for [vocational] work?" "You might as well put a technical-school teacher in a hospital to perform surgical operations," challenged Smith, "as to put a surgeon in charge of this work."[55]

Beyond this, however, vocational educators did not attack medicine per se, but rather the fact that all medical care under the surgeon general was *military* medical care. "It is the business of the War Department to fight," Prosser claimed, "not to give agricultural, commercial, and trade education for civilian occupations." The FBVE had statistics from France showing that it took an average of 12 to 18 months for maimed soldiers to recuperate, to receive instruction in a new trade, and to find employment. At the hearings, Prosser, Kidner, and McMurtrie all fervently argued that keeping disabled soldiers under military command beyond the point of physical restoration would hinder a man's ability to achieve occupational health. "Under military authority," contended McMurtrie, "many of the characteristics which make for success in civilian life . . . [are] stifled." Whereas the "soldier eats, sleeps, and works by schedule and rule," he continued, "the civilian must act on individual initiative and judgment."[56]

To the vocational educators, military control not only inhibited a man's ability to act as on his own initiative, but more crucially, it corroded the patient-healer relationship, making it impossible for soldiers to trust their caregivers. Drawing on his experience in Canada, Kidner told the congressional committee that it was necessary for vocational instructors to "wear civilian clothes and teach as civilians." "There is a moral reason for that," Kidner explained, "when the vocational instructor [was] in uniform, there was a certain amount of reserve between the soldier and the instructor." Kidner, Prosser, and McMurtrie all spoke of the need for a disabled man to "confide" in his vocational trainer. When choosing between a physician and a vocational educator, disabled men, claimed Prosser, more often went "to their civilian teacher and [told] him of their troubles."[57]

The strongest argument in favor of the FBVE taking over vocational rehabilitation rather than the OSG was cost. With Smith telling his fellow congressman that the expense of long hospital stays would be too great for the federal government to bear, Prosser sold vocational education as a "good business proposition." "Anything that can be done for a man through vocational rehabilitation," Prosser maintained, "that turns him from a wreck . . . a dependent and a social parasite into an orderly, happy, and successful wage-earning citizen will . . . increase the value of our human resources far beyond the cost of the training." Many congressmen appreciated this argument of conservation, especially in light of the ever-increasing Civil War pension roll. Indeed, to win support for the FBVE-sponsored bill, Prosser made a direct allusion to the pension burden. "I have seen man after man, who after the Civil War, would have been a dependent all his days on his meager pension," Prosser contented, "going out after a course of vocational rehabilitation to earn $75–150/month."[58] Smith employed similar rhetoric, maintaining that through vocational training disabled soldiers could be made "capable of becoming a part of the economic life of their country," as opposed to becoming deadweight pensioners, who, if left unattended, would "lapse into idleness and uselessness."[59]

Well aware of the immensely powerful role that politics played in the legislation of rehabilitation, Billings attributed the vocational educators' statements against the medical profession to the "jealousies existing between the various [governmental] departments," each of which had its own "axe to grind." Billings, who had practiced civilian medicine for over thirty-five years and had only recently donned a military uniform six months prior to the congressional hearings on rehabilitation, rebuffed the caricature of World War I physicians as inhumane, militaristic authorities. He pointed out that most of the men serving in the army's Medical Department were civilian practitioners who were called to serve, "to obey orders to go overseas."[60] In his Chicago days, Billings was known as an eager civil servant, and at the outbreak of the war he enthusiastically went to Washington at Gorgas's command, interrupting his work at the University of Chicago Medical School until after the war.

Billings took great offense at the vocational educators' ungratefulness toward military physicians—the first and only professionals to care for disabled US soldiers since the war broke out in April 1917. He pointed out that the Army Medical Department took charge of vocational education not only because of its expertise but also because in an emergency situation there was no other option. "We could not wait for legislation," Billings chided, "because men are disabled in camps today."[61]

Coming to Billings's defense, Senator James W. Wadsworth, Jr. (NY-Rep) pointed out that since the Army Medical Department had already begun vocational education programs, it would be redundant for the FBVE to provide the same services. Wadsworth visited a rehabilitation unit at Camp Fort McPherson in Georgia where he saw maimed soldiers being taught "therapeutic work in plumbing, tin-smithing, sheet-metal work, shoe repair, and telegraphing."[62] The strongest argument in favor of the FBVE taking over vocational rehabilitation rather than the OSG was cost. Concerned about fiscal expense—the FBVE was requesting a two million dollar annual budget—Wadsworth argued that the cost of vocational rehabilitation would be less if left in the Medical Department.[63] But his argument extended beyond economics. He believed that the medical profession bore most of the burden when it came to vocationally rehabilitating a disabled soldier. As he put it, "five-sixths of vocational education will be consumed by medical and surgical treatment," for the surgeon will always need to "be in consultation in fitting a hand to the injured man and readjusting it from time to time."[64] It was only fair, he thought, that the men who worked the most be given all federal monies and control.

Senator Joseph E. Ransdell (D-LA) concurred with Wadsworth.[65] Having personally visited Walter Reed General Hospital just several miles north of Capitol Hill, Ransdell explained that he had seen "soldiers actually being treated, being built over, so to speak; being remade, reconstructed." He had witnessed "limbless men who [had] made a success of life, rearing a family and taking good care of them." Why, Ransdell asked, "does not Senator [Smith] think that in the great hospitals, military and otherwise, there is an exceptional advantage for grouping the men together and providing vocational training for them which would not exist if you allow them to be discharged from those hospitals?"[66]

Both Ransdell and Billings worried that it would be more difficult to carry out vocational training once a maimed soldier had been discharged from the hospital, for vocational rehabilitation, as envisioned by the FBVE, was to take place in the individual's home or at the workplace. The FBVE planned to hire counselors who would travel from house to house, worksite to worksite, to provide onsite advice and training. Moreover, at the Interallied Rehabilitation Conference in Paris, French officials confessed that once patients had been discharged from medical services, it became very difficult to enforce vocational training. Indeed, only 20 percent of discharged French disabled soldiers who were recommended for vocational services complied. To avoid this problem, Billings believed that disabled soldiers should remain in military hospitals, for in the armed services a disabled man's most basic needs (food, lodging, clothing) could be met, allowing him to concentrate on developing his vocational abilities.[67] Senator Frank B. Brandegee (R-CT) agreed with Billings, contending that disabled soldiers should be "kept under the control of the Army where they could be *systematically* vocationally educated" (emphasis added).[68]

Billings understood the fears concerning militarism and worked hard to convince Congress that he and his colleagues were civilian physicians, first and foremost. He reminded Congress that the original OSG bill defined "military control" loosely, stating that physicians would not "dictate a man's training without considering his individual desires."[69] Defined this way, militarism was secondary to maintaining a trusting doctor-patient relationship. In Billings's words, "I wish the [surgeon general] could order me to do what I am doing without this military uniform," but by law the army general could not. Clothing, according to Billings, should not matter, for he believed that his and his colleagues' aim was true. "I speak for the Surgeon General, and I know for every man and every officer in the Division of Reconstruction," in saying that "we have not in mind anything but the man we want to take care of." In other words, despite their military pay, Billings believed that military rehabilitation physicians worked chiefly for the patient, not for the armed forces. "We are not clothed with military authority," he argued, "excepting as this uniform might give the appearance." Above all, he curtly concluded, "We are not martinets."[70]

Senator Jacob H. Gallinger (R-NH) made a final retort in defense of medicine, accusing the FBVE of mounting an "invasion" into medicine.[71] As a homeopathic physician and a senior congressman who earlier in his career had fought against the AMA (and Billings) to establish laws restricting animal experimentation, he was not always a friend of the medical establishment.[72] But contrary to animal experimentation legislation, this fight involved interprofessional, not intraprofessional, matters; the battle revolved around which profession—education or medicine—was best qualified to lead the federal program in rehabilitation. "Having had some knowledge of medicine and having a good deal of experience in hospitals," Gallinger argued, "I could not for the life of me . . . understand how [educators] . . . could . . . claim authority over the soldier."[73]

Another senator-physician, Joseph I. France (MD-Rep), accused Smith and the FBVE of employing an arcane definition of medicine in order to further their cause.[74] He reminded his opponents that "modern medicine" was not "only the knife of a surgeon or the drug prescribed by the physician, but [it was] also the recognition of exercise in the form of physiotherapy" and in the "from of work."[75] Here France echoed the argument of architect Edward Stevens, who in 1914 maintained that planners of the modern hospital had become too focused on the new operating theatres, neglecting the construction of facilities necessary for medical therapeutics. To solve the perceived imbalance, Stevens placed great emphasis on the need for mechanotherapy and hydrotherapy wards, departments that, as he put it, would "surpass the surgical operation building in size and equipment."[76]

In an attempt to give medicine the recognition he felt it deserved,

France motioned to change the name of the FBVE-sponsored bill from the "Vocational Rehabilitation Act" to the *"Physical Reconstruction and Vocational Rehabilitation Act."* The new name, France believed, would indicate that medical rehabilitation was essential to the recuperation of disabled men. Holding an immediate vote on the name-change in May 1917, the Senate rejected France's amendment.[77]

Despite opposition and criticism made in the name of medicine, the sixty-fifth Senate passed the FBVE-sponsored Federal Vocational Rehabilitation Bill on June 4, 1918, by a unanimous vote: 61 yeas and 0 nays. Opponents of the bill ended up casting a "yea" vote because they felt that their hands were tied. Since Smith presented the bill at the height of war, he was able to capitalize on wartime sentiment and patriotic duty. Although opposed to the FBVE gaining control over vocational rehabilitation, Senator Brandegee voted for it because he felt that he had "to try to do something to rehabilitate our wounded soldiers." He concluded with reservations, though, saying that he was "voting for a very dubious experiment." Similarly, Senator Gallinger confessed that he wanted to vote against the bill but after reconsideration decided to support it, for, as he put it, "I do not think my vote would be a determining factor one way or the other, as this measure seems to be the best we can get under existing conditions."[78]

The Federal Vocational Rehabilitation Bill did not get rid of veteran medical care. The OSG and the Division of Physical Reconstruction continued rehabilitating disabled soldiers, just as they had before the bill was passed. The only difference the bill made was in determining how vocational training would be delivered and who would be administering it. Under the act, hospitalized soldiers received intermittent visits from FBVE counselors who would continue to help the soldier find training and employment after he was discharged from the hospital.

The most crucial outcome of the Federal Vocational Rehabilitation Act was that vocational rehabilitation became disconnected from medical care, putting each mode of veteran (and eventually civilian) welfare on separate legislative tracks. Such a separation was seen as a triumph to the FBVE; by not giving control to the medical profession, vocational educators prevented rehabilitation from becoming fully medicalized. But in this case, demedicalization came at a great cost for disabled civilian workers. In September 1918, two months before the armistice, Smith introduced a bill granting vocational education to all disabled working-class men, but unlike the OSG version, Smith's bill was stripped of medical care.[79] Although the bill stalled in the Senate until the following year, the Civilian Vocational Rehabilitation Act, known as the Smith-Fess Act, passed both houses speedily in January 1920, making vocational rehabilitation a permanent federal law that covered all "persons disabled in industry or in any legitimate occupation."[80] Properly understood, then, the CVRA provided disabled workers with educational benefits, not health care. Ever fearful of expansive state powers, the United States has tended to prefer pedagogical solutions to social problems because, as historian Daniel T. Rodgers points out, they "generate so much less economically entrenched resistance than more direct measures."[81]

*   *   *

Just after stepping down as president of the AMA in 1906, Dr. Billings told a group of Chicago physicians that health insurance for the laboring class was "bound to come." "It does not make any difference what effect it will have on the medical profession," Billings maintained, "it is coming, and members of the medical profession should be willing to enter into an earnest investigation of the subject."[82] Over a decade later, while serving as the army's director of Physical Reconstruction, Billings had not changed his mind, even though by the time of the First World War a majority of his AMA colleagues strongly opposed all proposals of state medicine.

If Billings had gotten his way, the United States would have had a form of workers' health insurance, a system of governmentally funded rehabilitative health care for nonmilitary citizens. In the name of political expediency (and professional jealousies), vocational educators and the FBVE put an end to this possibility by making their work appear distinct and unrelated to a worker's physical well being. From the 1920s until the New Deal, US industrial workers would have to continue to secure medical care through fraternal societies and other private means, the same as they had prior to the war.[83] Rehabilitative medicine, though, would become much easier to find in post–World War I America. In the wake of the conflict, many private hospitals began offering physical and occupational therapy, seeing the additional therapies as a necessary "growth of social services."[84]

Billings's 1906 prediction ended up being half true. Government medicine did come to pass, but solely for military veterans, not for working-class civilians. One of the greatest legacies of World War I was the creation of the US Veteran's Bureau (later the Veterans Administration and now Veterans Affairs) in 1921, a department that not only consolidated the work of the War Risk Insurance Bureau, the US Public Health Service, and the rehabilitation division of the FBVE, but also instituted a federally funded system of hospitals devoted solely to the health care of America's soldiers.[85] Congress appropriated $18.6 million to the Treasury Department in March 1921 for the organization and construction of a centralized veterans' hospital system, where all forms of rehabilitative care could take place.[86] Billings served on the four-person physician committee that drew up the initial plans for the construction of the country's first veteran's hospitals.[87] By the end of the 1920s, after a decade of ambitious planning, construction, and millions of dollars in federal spending, 58 general and specialized hospitals had been built and placed under control of the US Veterans Bureau.[88] Today there are approximately 163 Veterans Administration hospitals (with at least one in each of the 48 contiguous states), treating over 4.2 million Americans yearly.

Although providing rehabilitative medical care to disabled soldiers was originally conceived of as a means to bring an end to the long-term commitment represented by veterans pensions, it ultimately resulted in a shifting of the nature and terms of the federal government's responsibility to its injured soldiers of war. The institution of rehabilitation during the First World War medicalized veteran welfare; it did not, as the original framers of the WRIA had hoped, lessen the economic burden of providing aid for disabled soldiers or eliminate the need to provide for veteran welfare. Indeed, in 1920 the US Treasury reported that it was spending approximately the same amount on World War I veterans as it was on Civil War pensioners.[89] The key difference between the two programs was how the money was allocated, with funds for World War I veterans going toward the construction of an extensive system of medical infrastructure instead of into the veterans' own pockets.

Walter Reed, Then and Now

True to its World War I legacy, Walter Reed has once again become the primary receiving hospital for injured American soldiers. Known as the "center of gravity" in army medicine, today's Walter Reed Army Medical Center (WRAMC) operates as a general hospital that offers a wide array of services equivalent to those found at any major university hospital in the civilian sector.[1] Like university hospitals, the WRAMC is known for its cutting-edge research as well as its areas of medical specialization. The pride of Walter Reed is—and always has been—its orthopedic and rehabilitative care. *Time* magazine reporter Michael Weisskopf, in Iraq to profile the American soldier as the "Person of the Year," underwent rehabilitation at Walter Reed after losing his hand in December 2003 from a grenade attack. He put it in the following way: "My fellow amputees . . . knew that if you had to lose a limb, you were in the right place, a citadel of excellence where President Eisenhower and generals from Pershing to MacArthur went to die."[2]

In a year or two, that history will come to an end. In May 2005, Defense Secretary Donald H. Rumsfeld placed Walter Reed on the Base Realignment and Closure Act (BRAC) list. According to Rumsfeld, closing Walter Reed was "essential" to winning the war. The century-old medical facility appeared outdated. Soldiers from Afghanistan and Iraq, Rumsfeld maintained, needed a "state-of-the art medical center."[3] Most important, rehabilitative care—especially services made available to amputee soldiers—needed to be brought into the twenty-first century.

As soon as Rumsfeld set base closures into motion, new construction began at the National Naval Medical Center in nearby Bethesda, Maryland, for a combined army-navy facility to open in 2012. Rather than improve the already existing structures, the closure plan will leave the original buildings at the Walter Reed campus fallow. Though the cost of relocation was estimated to be around $989 million, Rumsfeld maintained that it would end up saving the Department of Defense $145 million a year, once the new facilities open.[4]

Despite facing imminent closure, Walter Reed opened a state-of-the-art rehabilitation facility shortly after Rumsfeld's announcement, calling it the Military Advanced Training Center (MATC). For a place where medical miracles are said to happen, the center is housed in a rather unassuming building. Attached as an addition to the northwest corner of Walter Reed Hospital, a mammoth ten-story building with a perimeter equivalent to three football fields, the center pales in comparison. Measuring a mere 31,000 square feet, this simple two-story cinder-block building can be easily mistaken for a run-of-the-mill army barrack, an everyday structure for everyday soldierly life. But quite the opposite is true. For starters, the center houses a one-of-a-kind computer-assisted rehabilitation system, providing amputee soldiers with a whole host of simulated environments (e.g., jungles, forests, city streets) where they can test their new, state-of-the-art prosthetic limbs. It is here that today's military amputees are remade into fighting, active-duty soldiers. Although the center's external structure and title are ambiguous—the name, for instance, says nothing about medicine, rehabilitation, or amputees—the work that takes place inside is perfectly clear in its intent and vision.

While military rehabilitation officials laud the Military Advanced Training Center for its innovative technology and equipment, believing that such advances make for better and more comprehensive rehabilitative care, the facility is largely restricted to one goal and to one patient population: namely, the remasculinization of healthy, ambitious, amputee male soldiers. Embedded in the center's reason for being is the assumption that amputee men will be made manly again and that vocational and domestic rehabilitation are secondary pursuits to rehabilitative care.[5]

The fact that a rehabilitation facility can achieve only the most limited rehabilitation objectives for a limited patient population calls to mind author Mary Kaldor's provocative argument about the nation's military weapons systems. Those systems, she claimed, are not advanced; they are decadent, serving "as a kind of instant antidote to the troubles of the present epoch."[6] In a similar way, the center's rehabilitative technology, its rock-climbing walls, high-tech gait labs, and bionic limbs, serve as a salve for the terror and chaotic aftermath of war. What's more, money funneled into high-tech rehabilitation absorbs funds that could be used in other realms of patient care, as became obvious when the *Washington Post* sparked the "Walter Reed neglect scandal" by exposing a less attractive side of the hospital, with overworked case managers and dilapidated residential areas, and patient-soldiers living in rooms with rotting ceilings, mouse droppings, and cockroaches.[7] The new center dazzles the eye, but the underlying reality of war and its brutality persist, just out of view.

*   *   *

The Military Advanced Training Center opened on September 13, 2007, three-and-a-half years after the US-led invasion into Iraq. As is the case with most US declarations of war, aftercare was an afterthought in the nation's "Global War on Terror." One might think that as a country almost always involved in a military operation of one kind or another, the United States would be accustomed to and prepared for injured soldiers coming home. Yet this is not the case. Conflicts during the twenty years prior to Iraq (Grenada, the Gulf War, Somalia, and the Balkans) produced relatively few US causalities and thus never stressed the home front medical system as the current war has. Not since the Vietnam War has Walter Reed seen thousands of returning wounded soldiers, some needing months, if not years, of medical care.[8]

Those who sustained orthopedic injuries during the initial months of the Iraq war often found themselves admitted to Walter Reed's Ward 57, a unit devoted to amputee care. Prior to the war, Ward 57 had been populated with aging men and women receiving hip and knee replacements. Less than four months into the Iraq war, which began on March 20, 2003, Walter Reed had admitted 650 soldiers, filling Ward 57 to capacity with young amputee soldiers. Half of the soldiers arrived at Walter Reed after President Bush prematurely declared that major combat operations had come to an end in his "Mission Accomplished" speech on May 1.

The medical needs of this new patient population differed greatly from the aging population of veterans and their wives looking for joint replacements. Blast injuries—wounds produced by landmines, car bombs, and other improvised explosive devises—inflict uniquely gruesome and painful lesions. Land mines send shock waves up through the body, sheering soft tissue from the bone, slicing veins and arteries, and, at times, whiplashing the brain inside the skull. Military medical professionals speak of the uniqueness of battle injuries in terms of "co-morbidities." Plainly put, soldiers who sustain traumatic amputations (that is, limbs torn off by weapons of war rather than removed by a trained medical professional) can sometimes suffer from other medical conditions as well, such as traumatic brain injury, chronic back pain, shrapnel wounds, and post-traumatic stress disorder.

With this new patient population came a new ward culture—one defined by young men (a handful of women would arrive in the months and years to follow), barely beyond adolescence, coping with the reality of being permanently disabled. While these men shared a collective identity as the first group of war injured to return from Iraq, their adjustment to limblessness was highly individualized. Take the case of 1st Lt. John Fernandez. After graduating from West Point in 2000 with a degree in engineering, Fernandez joined the Third Battalion, Third Field Artillery in 2002 to, in his words, "help save the world from another tragedy like September 11th."[9] His battalion was among the first troops to move into Iraq. Two weeks into the war, on a reconnaissance mission 20 miles outside of Baghdad, Fernandez and his men came under attack by a laser-guided bomb, later discovered to have been launched by a US Air Force jet. Asleep on a cot next to his unit's Humvee at the time of the attack, Fernandez was blown to the ground. Upon zipping open his sleeping bag, he found that his feet were a "bloody pulp." Before attending to himself, he dragged himself back to the Humvee to haul the wounded gunner out of the vehicle before it blew up (the gunner later died). After a month in a Kuwaiti hospital, Fernandez arrived at Walter Reed's Ward 57 on April 11, 2003, with thick bandages at the end of both of his legs and an expectation of reconstructive surgery. By July, Fernandez had undergone twelve surgeries, leaving him a double amputee.[10]

Fernandez spent his time rehabilitating on Ward 57 with his 22-year-old wife, Kristi, always at his side. From the time he was admitted to Walter Reed, Fernandez insisted that he was "not going to feel sorry for" himself. The couple kept a positive attitude throughout Fernandez's stay, cracking jokes about how, with his one long, cone-shaped stump, he could dress up as a pirate for Halloween. To occupy themselves during large stretches of time between physical therapy, doctors visits, and prosthetic fittings, the newlyweds made plans for a formal wedding ceremony; they had had a quick, informal ceremony less than a month before Fernandez's deployment. After spending a couple months at Walter Reed, Kristi told a *Washington Post* reporter that, if someone had to get injured, she was glad it was her husband; she was convinced that together they were strong enough to face whatever challenges Fernandez's disability posed.[11]

Not all of the soldier-amputees on Ward 57 demonstrated the same kind of resolve to return peacefully to civilian life. Pfc Garth Stewart, for one, wanted to get back to Iraq as soon as possible. Stewart ended up in Walter Reed after he lost his left leg in a bunker explosion outside of Baghdad on April 5, 2003. Stewart did not take well to being hospitalized. His impatience with being a patient colored most of his interactions at Walter Reed. He nagged the attending surgeon, the nursing staff, and his physical therapist with the same, persistent question: "When can I get out?" Stewart went so far as to threaten to buy his own airline ticket back to Iraq. He missed the exhilaration of war and the feeling of being part of something larger than himself. "I mean, if someone came and got me out," he challenged his attending physician, "could the Army stop me from leaving?" Coming from Stewart, this was not an empty threat. With the Juan Ramón Jiménez adage "If they give you ruled paper, write the other way" tattooed across his chest, Stewart seemed willing to challenge authority.[12]

Little did Stewart know that his request to return to service would become a defining feature of amputee rehabilitation, a goal fully endorsed by President Bush himself. During a speech to the medical staff at Walter Reed on December 18, 2003, Bush remarked that "the medical care is so good and the recovery process is so technologically advanced," that an amputation should not hold back a soldier from returning to combat duty. After a morning visit with amputee patients undergoing physical and occupational therapy, he announced that forced discharges were a thing of the past. "Today if wounded service members want to remain in uniform," Bush declared, "the military tries to help them stay."[13]

In such an atmosphere of optimism and medical triumphalism, patients and caregivers alike believed that they were living through the dawning of a new age. Reflecting their confidence in the twenty-first-century army and its medicine, military health care providers self-consciously began to refer to soldier-amputees as "tactical athletes" rather than as patients.[14] Jeff Gambel, chief of Walter Reed's amputee clinic at the time, described the new generation of soldier-amputees in the following way: "Our guys and gals don't want to just walk household distances; they want to be able to return to running; they want to be able to return to duty." "And if they don't return to duty," Gambel added, "they want to be able to rock climb and do all those other things."[15]

For Gamble and his team of rehabilitators, the only thing that stood in the way of remaking these limbless men into tactical athletes was space and cutting-edge technology. By the one-year anniversary of Bush's declaration of victory, over 4,000 US soldiers had suffered causalities, hundreds of them amputees.[16] Military rehabilitation officials deemed Walter Reed's existing physical and occupational therapy gyms, spaces that housed more traditional, low-tech equipment found in most civilian hospital rehabilitation units, to be ill-equipped. Moderate-sized physical therapy gyms with conventional 10-foot-long parallel bars, wooden balance beams, and rocker boards (a simple wooden device used to improve a patient's balance) appeared incompatible with the technological goal of turning amputee soldiers into "bionic warriors," men who would become better, stronger, and faster through military rehabilitation.[17]

With no end to the war or its causalities in sight, in the summer of 2004 the rehabilitation team at Walter Reed put forth a plan for an "Amputee Advanced Training Center." During a congressional hearing that July, physical therapist Charles Scoville told the Committee on Veterans' Affairs that the Global War on Terror had produced more amputees relative to the overall number of wounded in action than any of the other major wars of the twentieth century. Whereas amputees accounted for 1.2–1.4 percent of all those wounded in action for both the world wars and the Korean War, Operation Iraqi Freedom and Operation Enduring Freedom had, by that point, produced twice as many amputees (2.4 percent) relative to those wounded in action. Technological advances in military protective gear—such as Kevlar vests that protect the torso (and thus vital organs) against bullets and grenade fragments—Scoville surmised, led to increased survival rates among the wounded. In previous wars, soldiers without such protection would have most likely died from blast injuries, due to internal bleeding. Whatever the cause for the increased number of amputees, the writing was on the wall. The war was producing men with missing limbs at a rate higher than ever before, and Walter Reed was responsible for getting them back on their feet again, with or without prosthetics.[18]

When the Walter Reed rehabilitation team put forth a formal proposal for a new amputee center, they relied not only on Scoville's testimony, but also on their experience of working within a hospital bureaucracy. The surest way to get the attention of hospital administrators and military bureaucrats was to raise the specter of fear concerning soldier-patient safety. According to one report, the cramped physical and occupational therapy gyms originally designed for "routine cases and workloads" were

WALTER REED, THEN AND NOW

"inadequate and unsafe for . . . advanced treatment." The limited amount of space, the report continued, created "hazardous situations for amputees, other patients, staff, and visitors."[19] In other words, the very same soldiers who narrowly escaped death in Iraq and Afghanistan were now being put in harm's way at home. While the language of "safety" and "hazardous situations" verges on the absurd in this context, the fact remains that US hospital administrators and military officials respond to reports of patient risk, especially in situations where legal suit can be brought against the institution. Health care institutions, especially those under the auspices of the federal government, are notoriously risk-adverse.

In addition to safety concerns, those agitating for a new amputee center insisted that Walter Reed needed to construct a place for "advanced training" in order to maintain its reputation as a leader in the field of physical rehabilitation. "Research, through observation and measurement of how the new prosthetics fit the needs of the patient, will . . . be hampered," the report reasoned, "by the lack of available space for equipment and staff to observe and measure."[20] If such experimentation stalled, the report concluded, effective improvements in prosthetic design would come to an end, making the goal of returning soldiers to active duty untenable.

Ever since World War I, when the original Limb Lab was first constructed, research in limb design has held a privileged place in military rehabilitation. Many branches of the Department of Defense have a stake in prosthetic design. During World War I, Woodrow Wilson's 1916 National Defense Act made prosthetic design a separate research program to be carried out with the help of the National Council of Research and under the direction of the army's Office of the Surgeon General. This trend continued into the Second World War and became an essential component of the Cold War. President Harry S. Truman's newly established National Security Council of 1947 married prosthetic design to the military-industrial complex, establishing key universities—Case Institute of Technology, Massachusetts Institute of Technology, Michigan State University, New York University, University of California at Los Angeles, and Western Reserve University—as sites dedicated to the advancement of prosthetic technology.[21] In the heat of the space race, President Dwight D. Eisenhower in 1958 created not only the National Aeronautics and Space Administration but also the Advanced Research Projects Agency (Defense was added in 1972).[22]

The Defense Advanced Research Projects Agency (DARPA) was established and exists today for the purpose of maintaining technological superiority over other nations. Both the Internet and stealth technology came out of its initiatives. In the early years, it funded university-based engineers to create "lifelike" limbs, applying the newest principles of biomechanics to the latest materials of Plexiglas, Lucite, polyester, silicone, titanium, and electronic servomotors. As historian David Serlin tells us, the first myoelectric arm emerged out of this military-university complex. In the early 1960s, MIT mathematician Norbert Wiener created a cybernetic arm, dubbed, like its World War I leg predecessor, "The Liberty Limb."[23]

Today, DARPA enjoys a $3 billion budget and acts, as author Michael Belfiore puts it, "like a small company that gets innovative technology projects funded quickly and with a minimum of bureaucratic hassle."[24] In addition to weapons design, the agency is still committed to repairing the very limbs that get blown off by the armament it helps to create. In his 2004 Congressional Statement of the House Committee on Veterans Affairs, Dr. Brett Giroir, the agency's deputy director testified that the singular goal of his agency was at once to engineer bombs and the bodies maimed by those bombs. "Our vision for amputee care," Giroir explained, "came directly out of [the] broad effort to harness insight from biology to make US war fighters and their equipment safer, stronger, and more effective." The researchers involved with the agency pride themselves on turning far-fetched ideas into reality, science fiction into everyday fact. For amputees coming back from Iraq, the agency holds out the promise of creating "biologically integrated, fully functional limb replacements that have *normal* sensory abilities" so that patients wearing prosthetic limbs can "sense an artificial limb's position without looking at it, and . . . 'feel' precisely what the artificial limb is touching."[25]

This vision has yet to become a reality. Prosthetic limbs such as those described by Giroir are primarily targeted at soldiers who need arm and hand replacements; the primary goal of leg replacements is weight-bearing, not sensory touch. Much like arm-amputees who refused to wear mechanized limbs developed during the First World War, arm-amputees from the Iraq War are also deciding to wear more simple prosthetic arms because the high-tech arms are too complicated to use and too uncomfortable to wear.[26] More often than not, high-tech arms hinder rather than assist their users.[27] Take, for example, Weisskopf's description of receiving his first myoelectric arm: "The prosthesis made my arm crook out like Popeye's; my range of motion was so limited that I couldn't raise the hand within a foot of my mouth. I kept bumping it into things. . . . I named it Ralph, after the clumsiest kid in my grade school."[28]

Cautionary tales from the past or present, though, are rarely acknowledged or voiced among military rehabilitation officials, for the agency holds out the hope that through technological advancement the United States can magically mend the wounds of war. Giroir said as much in his testimony before Congress. The goal, he boasted, "is for amputees to return to normal life, with no limits whatsoever."[29] If the agency could wipe away disability, then the goal of returning soldier-amputees to combat duty seemed rather simple.

The Walter Reed rehabilitation officials' arguments that better technology and space were needed for better care proved persuasive.[30] Congress moved swiftly to approve $10 million for the design and construction of the Amputee Patient Care Program at Walter Reed, plus $1.4 million to equip the center. Less than three months after Scoville's and Giroir's testimonies, Walter Reed held a "kickoff" meeting for the center and, by November 2004, a ceremonial groundbreaking. With everything moving at such a fast clip, the Army Corps of Engineers (the branch responsible for new construction) predicted that the project would be completed by April 2006.

The architectural firm Smith Group drew up the first plan that the Army Corps of Engineers considered. The firm had a long history of working with the Department of Defense and had just completed a new $100 million expansion to the Defense Intelligence Analysis Center on Boling Air Force Base in Washington. The Smith Group's plan for the Amputee Center was to create a three-story, L-shaped building, topped off by an all-glass breezeway that connected the new center to Walter Reed's main medical center building. The breezeway would offer access to a green space roof with a healing garden area, a design feature that not only "harmonized" the new structure with the original hospital but also improved patient care. The green space roof, however, pushed the project over budget, and the Smith Group refused to make the necessary changes to decrease costs. Six months after the November groundbreaking, the Smith Group pulled out of negotiations.

Before the Army Corps of Engineers could even review a competing proposal, all plans for the Amputee Center came to a halt. Rumsfeld's announcement of base closings put the construction of a new rehabilitation facility into question. Yet, just as the department put Walter Reed on the closings list, the hospital experienced the heaviest influx of injured soldiers since the outbreak of war, with an estimated 1,000 casualties admitted every month.[31] Giant C-17 "medevac" jets landed almost every night at nearby Andrews Air Force Base. There, oversized ambulances (i.e., white buses converted into emergency medical vehicles) waited to transport newly injured soldiers, many with blood and sand still under their nails, to Walter Reed.

Because of the ever increasing number of amputees, the rehabilitation team continued to persist in getting the Amputee Center built, despite Rumsfeld bringing an end to Walter Reed's one hundred year history. Finally in April 2006, when, according to the original timeline, the Amputee Center was supposed to be completed, Congress reapproved the project with the stipulation that the center be built as a "temporary transitional structure," a building that could be moved to Bethesda according to the closings guidelines. Since the Smith Group went over budget on the original proposal, their competitor, the architectural firm of Ellerbe Becket won the contract.

Tom Anglim, the principal architect representing Ellerbe Becket, knew his clients did not want a run-of-the-mill medical facility. The purpose of the building, after all, was to rehabilitate "tactical athletes," soldiers who wanted to return to a highly active lifestyle, whether as combat soldiers or rock climbers. Anglim thus designed the building to, in his words, "create the feeling of a contemporary athletic training facility rather than the feeling of being in a hospital."[32] The aim of creating a health care space that looked like an athletic training area was an easy assignment for Ellerbe Becket. In addition to a long resume of designing major sport arenas (e.g., the Verizon Center in Washington, DC, renovations to Madison Square Garden in New York, and Yale's Payne Whitney Gym renovations), the firm was also responsible for renovating some of the country's leading medical facilities, such as the Mayo Clinic and the Shriners Hospitals for Children.

The new center, christened the Military Advanced Training Center, opened to great military fanfare. While Anglim admitted it was "not a building you'll see on the cover of Architectural Record," military officials and soldiers voiced nothing but praise for the center's innovativeness. In addition to a 225-foot indoor track with the world's first oval support harness system for safety purposes, the center became one of the first in the world to house a highly specialized computer-assisted rehabilitative environment. The system is akin to a 3-D virtual reality video game, with a floor-to-ceiling curved projection screen and a platform that can mimic everything from a jungle floor to a snowy slope. (The platform is, in reality, a treadmill bolted to a helicopter simulator.) The "speed boat" program, for instance, makes patients feel as if they are standing on the hull of a watercraft, as the platform swerves and undulates to the rhythm of oncoming waves. Patients direct the boat through a race course of buoy gates—much like a giant slalom race—shifting their weight side-to-side and back-and-forth. One of the other popular programs is the city street simulation. As Scoville describes it: "We can start with the patient walking on an empty street and gradually add parked cars, traffic, pedestrians, and noise. We'll take the patient to the edge of discomfort, but not beyond what they [*sic*] can handle."[33]

In addition to the 3-D system, the center's Gait Lab received high acclaim on opening day. The lab sits on a large isolated floating slab of concrete, a feature necessary to keep out vibrations (and thus corrupted data) from neighboring rooms. The room is equipped with 23 motion-capture cameras (the type commonly used in the movie industry for animation), six force plates (to register how patients carry their weight as they walk), and a sunken treadmill, flush to the floor. The 2-D views captured from the multiple cameras feed into a computer software system that produces a single, composite 3-D image of an individual's gait pattern, which is then used to help determine how a user's residual limb interacts with his prosthetic device.[34]

The facilities that mimic combat, however, are among the most popular features of the center. Along with a 22-foot rock wall, the center offers a military vehicle simulator as well as a firearms training simulator. For Ramon Padilla, an amputee patient who was invited to try out the center on opening day, the weapon simulator was the highlight of the place. As soon as Padilla pulled the trigger of a modified M4 with the hook of his prosthetic arm, he exclaimed "Awesome!" When asked what he thought about the millions of dollars that went into the MATC, Padilla told one *Washington Post* reporter, "Screw the expenses. Do what you have to do to help soldiers recover better and to have a healthy life."[35]

Padilla has a point. In the scheme of the Department of Defense's multitrillion dollar spending, $10 million is miniscule. In 2005 alone, the defense budget was over $400 billion, with $104 billion going toward military personnel salaries and $70 billion to weapons research and development, most notably to the development of unmanned aerial vehicles, known as the Joint Strike Fighter. The Defense Health Program, by contrast, received only $18.2 billion, with $19.2 million going toward amputee care at Walter Reed.[36]

Seen in this light, the Military Advanced Training Center is a low-cost, high-payback endeavor. Everyone involved—from congressmen, rehabilitation workers, and soldier-patients to the public—is made to feel good. Even the workers employed by Turner Construction Company who laid the actual bricks and mortar of the building felt uniquely inspired by the project. Seeing the injured troops on a daily basis made them work faster—so fast that the building was finished ahead of schedule. "Everybody on the project," Elihu P. Hirsch, project manager, claimed, worked "with great feelings of satisfaction."[37]

This outpouring of public support for the care of disabled soldiers, especially amputee soldiers, is commonly seen among nations at war, past and present. There is a debt to be paid. The soldier who puts his life in harm's way and comes out injured deserves some sort of recompense. From the time of the Civil War to the eve of the First World War, the United States compensated its injured citizen-soldiers with pensions. Disabled soldiers were given enough money to eke out a living without having to find gainful employment. Come World War I, the United States radically changed course and began repaying disabled citizen-soldiers through a complex system of war risk insurance, physical rehabilitation, and vocational retraining—a medicalized system of aftercare set up to get soldiers back to the industrial workplace and "off the dole."

The training center is an extension of this World War I vision of compensation, but with some important differences. Whereas the goal of World War I rehabilitation was to move disabled soldiers from the "military Army to the industrial Army," to get them integrated into the industrial workplace, today's rehabilitation intends to turn out "tactical athletes" and combat soldiers. This change arises not from a shift in rehabilitation practice, but rather from the nature of military service itself. Because the US military now relies on an all-volunteer army, the soldiers who end up injured at Walter Reed are mostly career soldiers.[38] To get them back to their job is to return them to combat duty.

Yet on-the-ground practice shows that rehabilitation is much messier than it is often portrayed. For every success story that comes out of the training center, there are hundreds of other injured soldiers for whom the goal of participating once again in extreme sports is completely untenable. Certain soldier-patients who suffer from Traumatic Brain Injury, a diagnostic group that, by some estimates, accounts for as much as 70 percent of all causalities from the wars in Iraq and Afghanistan, are so neurologically impaired that simple activities of daily living become akin to the performance of extreme sports for an able-bodied person.[39] The problem with military physical rehabilitation, then and now, is that it is one-size-fits-all, with male amputee care as the model.

The goal, then, of returning these injured tactical athletes to their former selves is often unrealistic, even for those who are neurologically intact and have the boundless motivation to get there. Stewart, the amputee on Ward 57 who threatened to buy his own private airline ticket back to Iraq, did not achieve his goal of rejoining his unit, despite his monumental efforts. After undergoing rehab at Walter Reed for several months, Stewart returned to Fort Benning, where he completed "hand-to-hand combat school and unloaded dozens of 100-pound ammo boxes in field exercises," all with the use of a prosthetic leg.[40] But Stewart could not meet the demands of running long distances, because his stump would blister and swell, making artificial limb wear difficult, if not impossible. Stewart eventually retired from the service and in 2005 went on to become an undergraduate at Columbia University, majoring in history.[41]

No group challenges the ethic of today's military rehabilitation effort more than the small cohort of female combat amputees who have returned from Iraq and Afghanistan. Juanita Wilson, the war's fourth female combat amputee, arrived at Walter Reed on August 25, 2004, with a severed left arm. From the time she got to Walter Reed her primary concern was to protect her 6-year-old-daughter, Kenya, from the truth of her injury. In the beginning Wilson went to great lengths to hide her disability from her daughter. On their first outing together, Wilson was sure to have the "nurse put makeup on her face" and "stow her IV medications into a backpack."[42] When Wilson became an outpatient at Walter Reed, she quickly learned that she could not keep up the façade. As soon as Kenya asked her mom to make her a sandwich, the challenges that come with being a one-armed mother set in. She could not longer hide the reality of her physical condition from herself or her daughter.

Female combat amputee Dawn Halfaker told a *Washington Post* reporter that to be a woman with an amputated limb is "not better or worse, just different" than it is for a man. She speculated that for some men, a combat amputation is a "badge of honor that they do not mind showing [off]." "For a woman, at least for me," she explained, "it's not at all [like that]." "I only have one arm, I'm okay with that," she continued, "but I want to be able to walk around and look like everyone else and not attract attention to myself." Although Halfaker was an elite athlete upon entering service and a standout basketball player at West Point, she had to retire from the army as captain after she lost her right arm in the war. She still plays sports, but with one arm, since she found mechanized prosthetics to be of little help in her everyday life. When she wears an artificial limb, a light-weight, natural looking replica of a flesh and blood arm, it is to "blend in with the outside world."[43]

To be sure, the Military Advanced Training Center offers a host of psychological and occupational therapy services that assist patients with adapting to their individual circumstances; Wilson no doubt learned how to make a sandwich with (or without) the assistance of her prosthetic arm. But these are not the stories that make headlines, nor are these the rehabilitation goals around which the center has been constructed. Female amputees coming back from Iraq and Afghanistan do not seem to be concerned about becoming tactical athletes. Rather, most want to return to their lives as mothers and civilians, goals that are, in many ways, more complex and difficult to achieve than mere athleticism.

The center was built for male patients like Captain David M. Rozelle. While driving a Humvee toward Hit, Iraq, in June 2004, Rozelle ran over an antitank mine, causing a traumatic amputation of his right foot and ankle. After nine months of rehabilitation at Walter Reed and in Breckenridge, Colorado (where he participated in the National Disabled Veterans Winter Sports Clinic), Rozelle returned to Iraq in April 2005 as commander of the 3rd Armored Cavalry Regiment. He was the first combat amputee to return to the zone of war.[44] Rozelle was injured too early in the war to benefit from the advanced training center, but he was instrumental in its construction. Indeed, just by virtue of being the first amputee who returned to combat, Rozelle earned an advisory position on the design and construction of the facility.[45]

Exceptional soldier-amputees such as Rozelle keep the hope alive that medicine can cure disability and, more important, solve the social problems brought about by a war that fundamentally disrupts the lives of its citizens. This is the dream that motivated World War I orthopedic surgeon Dr. David Silver, chief of the Walter Reed Limb Lab, and his insistence that the E-Z-Limb become standard issue for all lower extremity amputees. In Silver's mind, the E-Z-Limb, with its lifelike looks, made an amputee's disability disappear. With an E-Z-Limb, a (once) disabled man could easily land a job and a wife, the two accomplishments that would make him normal again, in both the public and the private spheres. The same logic holds today. David Polly, former chief of orthopedic surgery at Walter Reed, told a reporter in 2003 that while "young, virile males" have some difficulties adjusting to the amputation of a limb, they can be assured that through the powers of modern medicine and biotechnology they will "still get a date . . . and still be a husband."[46]

In a sense, exceptional amputee patients (or what are known as "supercrips" among disability studies scholars) have always defined military rehabilitation's public image and its practice. The army's World War I propaganda journal *Carry On: A Magazine on the Reconstruction of Disabled Soldiers and Sailors* (1918–19) photographed amputee soldiers who "made good," that is, who kept busy at work despite missing a limb, while regularly passing over less photogenic soldiers with tuberculosis, facial disfigurements, and crushed skulls. The text of the magazine reflected the same bias. The more complex conditions and the so-called invisible diseases, such as shell shock, deafness, and blindness, put a damper on the explicit and self-consciously cheerful tone of *Carry On*.

The same normative assumptions permeate the thinking of DARPA and rehabilitation officials today. Nothing satisfies the human eye or quells the pangs of guilt from seeing a war-torn body like seeing an amputee who has been made to appear whole by an artificial limb, especially one that looks (or better, acts) like the real thing. That President Bush was photographed time-and-again on the White House grounds jogging next to amputee soldiers, outfitted in C-legs, was no accident. A dismembered body that can be put back together again, and work as it did before, creates the momentary illusion that there is no human cost of war—that there is no "waste" in war.

Today Michael Weisskopf wears a hook arm to report and write stories for *Time* magazine. He cast away his lifelike prosthetic hand because it required too much maintenance and offered little functionality. He chose an age-old technology because the cutting-edge prosthesis afforded him "the precision of a boxing glove." When he finally opted for the hook arm over the more cosmetic, normal-looking one almost a year after he sustained his amputation, he wondered why he had tried so hard to pass as able bodied during his first year of recovery. "Why try to conceal a handicap?" he asked.[47] Why? Because the ethic of rehabilitation, established almost a century ago, lives on.

Acknowledgments

It gives me great satisfaction to thank the many people who have helped me bring this book to fruition. Many thanks go to Michael G. Rhode who, on repeated occasions, willingly shared his in-depth knowledge of the material held at the National Museum of Health and Medicine. Hunting down sources in the National Archives and Records Administration (NARA) in College Park, Maryland, would have taken me far longer if not for the help of Mitchell Yockelson, who knows the World War I holdings inside and out. Because of Mitch, I also had the pleasure of working at the NARA in Washington, DC, with Trevor K. Plante, who helped me negotiate the materials of the Records of the Department of Veterans Affairs. Special thanks go to Richard Boylan, Senior Military Reference Archivist at the NARA in College Park who went above and beyond the call of duty in helping me locate source material—namely, Inspector General Reports—that provided the disabled soldier's perspective. I am also grateful to countless librarians and archivists at Carlisle Barracks, the College of Physicians of Philadelphia, the New York Academy of Medicine (especially Christopher Warren and Arlene M. Tuchman), and the American Physical Therapy Association. Nick Okrent at the University of Pennsylvania Library and Toby Appel at Yale's Cushing/Whitney Medical Historical Library were especially helpful.

Research trips to the Washington, DC, area and beyond would not have been possible without generous financial support. This project received its first material backing from the Graduate School, History Department, and Program in the History of Science and Medicine at Yale University.

Additional funding came from the Donaghue Initiative in Biomedical and Behavioral Research Ethics and the Wood Institute Research Fellowship of the College of Physicians of Philadelphia. As a faculty member at the University of Pennsylvania, I have benefited from the Robert Wood Johnson Health and Society Scholars Program, the Trustee's Council of Penn Women, and the Penn Humanities Forum.

Research for this book began at Yale University where I had the good fortune of working with some of the best historians in North America. I am greatly indebted to John Harley Warner, who has been a constant counsel and an intellectual inspiration at every step of the way. From him I learned the essence of the historical trade: to create a comprehensible narrative that takes account of the complexity and multiplicity of historical events and ideas. His unfailing dedication to my intellectual and professional development has been more important to my becoming a historian than he can ever know. Glenda E. Gilmore, with her sharp wit and insights, kept me on my toes, urging me never to be satisfied with a superficial read of the past, or a flimsy argument. Daniel J. Kevles has been a steadfast supporter of both my professional development and research pursuits. Susan Lederer greatly enriched my understanding of medical and scientific practices during the early twentieth century, through both her written work and encyclopedic mastery of primary and secondary historical source material. Naomi Rogers, who unwittingly set me off on my research path, deserves special credit not only for inspiring me to write about disabled soldiers, but also for redirecting me when I faced roadblocks along the way. Her uncanny ability to see to the very core of my project made her an invaluable critic and colleague. And finally, many thanks go to Roger Cooter for comments that unfailingly provided a wholly unique perspective on things.

My colleagues and graduate students at the University of Pennsylvania have made research and writing a joy, providing me with a scholarly community strong both in its intellectual vitality and camaraderie. Rosemary A. Stevens and Patricia D'Antonio graciously read the entire manuscript with the utmost care and precision. The members of the Department of the History and Sociology of Science—Mark Adams, Robert Aronowitz, David Barnes, Ruth Cowan, Nathan Ensmenger, Steven Feierman, Ann Greene, Robert Kohler, Henrika Kuklick, Susan Lindee, Jonathan Moreno, and John Tresch—have all lent me their support and provided guidance at crucial steps along the way. Thanks also to Kathy Brown, Cynthia Connolly, Ruth Cowan, Julie Fairman, Susan Lindee, Nancy Hirschmann, Kathy Peiss, and Audra Wolfe for their continual professional guidance and friendship. Meghan Crnic, Jessica Martucci, and Samantha Muka diligently and adeptly assisted me in the final stages of putting the manuscript together. Patricia Johnson made coming to the office a happy experience.

Several friends and colleagues have at various stages read portions of this book, making it a much stronger piece of scholarship. On this count I am indebted to Cynthia Connolly, Rebecca Davis, Maureen Flanagan, Mark Krasovic, Sally Romano, and Dominique Tobbell for taking the time out from their busy lives to provide much-needed criticism and advice. I also benefited immensely from the feedback I received while presenting sections of the book to the Johns Hopkins University Institute of the History of Medicine, the Department of the Social Studies of Medicine at McGill University, the Colloquium on the History of Science, Technology, and Environment at the University of Maryland, the Washington Society for the History of Medicine, the New York Academy of Medicine, the College of Physicians of Philadelphia, the National Museum of Health and Medicine, the Penn Humanities Forum, and the Barbara Bates Center for the History of Nursing at the University of Pennsylvania. To each and every colleague who has listened to and commented on my presentations, I thank you. I was lucky enough to have my book land in the hands of Karen Darling, who has proven to be an ideal editor; she met every need that I (and the book) ever had with enthusiasm and aplomb. The anonymous readers of the manuscript put great energies into making this book a better read and a sturdier history. Thomas Schlich deserves extra credit for having reviewed the manuscript twice with supreme intellectual rigor and thoroughness.

I never would have been able to finish this project without the loving support of my mother, Rose Anne O'Donnell, my siblings, and the dedicated women who cared for my two young children while I was away at work. Special thanks goes to Lynne Norris, Maureen Fontaine, Diana Gollnest, Frederike Werner, Sadie Brumfield, and the teachers at Overbrook Preschool and Kindergarten for nurturing and educating my children.

Loving thanks to Kaitlyn and Mark who made me laugh and smile along the way, teaching me not to take life so seriously. To my husband, Damon Linker, who has been my number one supporter through thick and thin, I dedicate this book. He believed in my abilities before I had even recognized them myself. I have benefited enormously from his cheerful willingness to read (and re-read) every page of this book, his formidable intellect, his ear for good writing, his passion for learning and ideas, and, last but not least, his companionship.

Notes

INTRODUCTION

1. One can view a video clip of this scene at http://www.front
   runnermagazine.com/2009/10/08/rest-when-youre-dead-
   the-story-of-garth-stewart/; Matt Mireles, "Garth Stewart:
   The Untroubled Soldier," Matt Mireles Blog, http://www
   .mattmireles.com/Garth%20Stewart.html/; "Rest When
   You're Dead: The Story of Garth Stewart," Frontrunner Magazine, http://www.frontrunnermagazine.com/2009/10/08/
   rest-when-youre-dead-the-story-of-garth-stewart/.
2. Ibid.
3. For an overview of the development of the US veterans' pension system, see the President's Commission on Veterans'
   Pensions, *The Historical Development of Veterans' Benefit in the
   United States* (Washington, DC: GPO, 1956). For the Revolutionary War, see John Resch, *Suffering Soldiers: Revolutionary
   War Veterans, Moral Sentiment, and Political Culture in the
   Early Republic* (Amherst: University of Massachusetts Press,
   1999), and "Politics and Public Culture: The Revolutionary
   War Pension Act of 1818," *Journal of the Early Republic* 8
   (Summer 1988): 139–58. For the Civil War pension system,
   see Theda Skocpol, *Protecting Soldiers and Mothers: The Politi-
   cal Origins of Social Policy in the United States* (Cambridge, MA:
   Harvard University Press, 1992) as well as Larry Logue and
   Peter Blanck, *Race, Ethnicity, and Disability: Veterans and
   Benefits in Post–Civil War America* (Cambridge: Cambridge
   University Press, 2010).
4. Larry M. Logue, "Union Veterans and Their Government:
   The Effects of Public Policies on Private Lives," *Journal of
   Interdisciplinary History* 22 (Winter 1992): 412.

5. As quoted in John William Oliver, "History of the Civil War Military Pensions, 1861–1885," *Bulletin of the University of Wisconsin*, no. 844, History Series, no. 1 (1917): 19.

6. Resch discusses the linkage between pensions and national gratitude in *Suffering Soldiers*.

7. While this book traces a significant change in US veteran welfare policy, it also tells a story of continuity. In her thorough account of the Civil War pension system, political scientist Theda Skocpol argues that public provision for veterans "died with the Civil War generation, and was not replaced with other measures until the Great Depression and the New Deal of the 1930s" (*Protecting Soldiers and Mothers*, 2). In actuality, public provisions for veterans never ended; they just changed in nature and character. This book demonstrates that, contrary to Skocpol's periodization, New Deal veteran welfare policy grew out of legislation passed during the Progressive Era. The laws that established rehabilitation for disabled soldiers of the First World War, for example, led in 1921 to the creation of the Veteran's Bureau, later the Veterans Administration and now the Department of Veterans Affairs, which manages hundreds of hospitals across the country, providing federally funded health care to all veterans of war. Rehabilitation was (and still is) a form of medical and educational welfare that aims to ameliorate veteran demands for compensation while also lessening the cost of pension and disability payments.

8. Progressive reformers commonly used the words "cripple" and "handicap" to describe disabilities. While this book is mindful that disability advocates today might find this kind of language offensive, it nevertheless utilizes the terminology of its historical actors in order to more accurately reflect the meaning of the rehabilitation movement. To see an example of disability rhetoric from the World War I era, see Douglas McMurtrie, *The Organization, Work, and Method of the Red Cross Institute for Crippled and Disabled Men* (New York: Red Cross Institute for Crippled and Disabled Men, 1918).

9. "Science to Rebuild Our War Cripples," *New York Times*, September 17, 1917, 6.

10. For the best surveys of US involvement in the First World War, see David M. Kennedy, *Over Here: The First World War and American Society* (New York: Oxford University Press, 1980), and Jennifer D. Keene, *Doughboys, the Great War, and the Remaking of America* (Baltimore: Johns Hopkins University Press, 2001).

11. For numbers and figures see Skocpol, *Protecting Soldiers and Mothers*, 109–10, and William H. Glasson, *Federal Military Pensions in the United States* (New York: Oxford University Press, 1918), esp. chap. 3 and the statistical tables found on p. 273.

12. For more on the ideological bent of the rehabilitation movement, see "Crippled Soldiers Will Not Be Paupers," *New York Times*, August 13, 1917, 11; and "Ask Health Camps for Unfit of Nation: E. E. Rittenhouse Wants

Government to Help Correct Defects of Manhood," *New York Times,* May 13, 1917, E1.

13. For "ugly law" legislation and the practice of eugenic sterilization, see Susan M. Schweik, *The Ugly Laws: Disability in Public* (New York: New York University, 2009), and Martin S. Pernick, *The Black Stork: Eugenics and the Death of "Defective" Babies in American Medicine and Motion Pictures since 1915* (New York: Oxford University Press, 1996).

14. For an example of how rehabilitation officials believed work to be redemptive, see Garrard Harris, *The Redemption of the Disabled: A Study of Programmes of Rehabilitation for the Disabled of War and Industry* (New York: D. Appleton, 1919). Subsequent versions of this book were published in Canada under the same title, but with Dr. Frank Billings—a figure who will be discussed at length later in the book—as a co-author. For literature on the importance of work and the work ethic to US Progressives, see Daniel T. Rodgers, *The Work Ethic in Industrial America, 1850–1920* (Chicago: University of Chicago Press, 1978), and Michael Katz, *In the Shadow of the Poorhouse: A Social History of Welfare in America* (New York: Basic Books, 1986).

15. As quoted in Kim E. Nielsen, *The Radical Lives of Helen Keller* (New York: New York University Press, 2004), 28.

16. For the history of the use of work therapy in medicine, see Gerald N. Grob, *The Mad among Us: A History of the Care of America's Mentally Ill* (New York: Free Press, 1994); Grob, *From Asylum to Community: Mental Health Policy in Modern America* (Princeton, NJ: Princeton University Press, 1991); and David J. Rothman, *The Discovery of the Asylum: Social Order and Disorder in the New Republic* (Boston: Little, Brown, 1971).

17. Mark Aldrich, *Safety First: Technology, Labor, and Business in the Building of American Work Safety, 1870–1939* (Baltimore: Johns Hopkins University Press, 1997).

18. Lawrence B. Glickman, *A Living Wage: American Workers and the Making of Consumer Society* (Ithaca, NY: Cornell University Press, 1997).

19. Emphasis in the original. The Creed appears in multiple issues of *Carry On: A Magazine on the Reconstruction of Disabled Soldiers and Sailors* and the *Publications of the Red Cross Institute for Crippled and Disabled Men* series. For a copy of the Creed, see *Carry On* 1 (June 1919): 1.

20. Daniel T. Rodgers, *Atlantic Crossings: Social Politics in a Progressive Age* (Cambridge, MA: Harvard University Press, 1998).

21. The earliest accounts of rehabilitation in Europe include Robert Gerald Whalen, *Bitter Wounds: German Victims of the Great War, 1914–1939* (Ithaca, NY: Cornell University Press, 1984); Roger Cooter, *Surgery and Society in Peace and War: Orthopaedics and the Organization of Modern Medicine, 1880–1948* (London: Macmillan Press, 1993); and Seth Koven, "Remembering and Dismemberment: Crippled Children, Wounded Soldiers, and the Great War in Great Britain," *American Historical Review* (1994): 1167–1202.

For recent works on World War I disabled veterans, see Thomas Schlich, "The Perfect Machine: Lorenz Böhler's Rationalized Fracture Treatment in WWI," *Isis* 100 (2009): 758–91; Jeffrey Reznick, *Healing the Nation: Soldiers and the Culture of Caregiving in Britain during the Great War* (Manchester: Manchester University Press, 2004); Deborah Cohen, *The War Come Home: Disabled Veterans in Britain and Germany, 1914–1939* (Berkeley: University of California Press, 2001); Ana Carden-Coyne, *Reconstructing the Body: Classicism, Modernism, and the First World War* (Oxford: Oxford University Press, 2009); Roxanne Panchasi, "Reconstructions: Prosthetics and the Rehabilitation of the Male Body in World War I France," *Differences: A Journal of Feminist Cultural Studies* 7, no. 3 (1995): 109–40; Heather R. Perry, "Recycling the Disabled: Army, Medicine, and Society in World War I Germany" (PhD diss., Indiana University, 2005); and Joanna Bourke, *Dismembering the Male: Men's Bodies, Britain and the Great War* (London: Reaktion Books, 1996). For a post-structuralist critique of the rise of rehabilitation, with an emphasis on France, see Henri-Jacques Striker, *A History of Disability*, trans. William Sayers (Ann Arbor: University of Michigan Press, 1999), esp. chap. 6.

22. The literature concerning rehabilitation in non-European nations is scant. There are but a few exceptions. For Australia, see Marina Larsson, *Shattered Anzacs: Living with the Scars of War* (Sydney, Australia: University of New South Wales Press, 2009). For Canada, see Desmond Morton and Glenn Wright, *Winning the Second Battle: Canadian Veterans and the Return to Civilian Life, 1915–1930* (Toronto: University of Toronto Press, 1987). Ana Carden-Coyne has written an article-length piece about US disabled veterans from the First World War, "Ungrateful Bodies: Rehabilitation, Resistance and Disabled American Veterans of the First World War," *European Review of History* 14, no. 4 (2007): 543–65. An account of India still remains to be written. Mark Harrison offers a good starting point in "Disease, Discipline and Dissent: The Indian Army in France and England, 1914–1915," in Roger Cooter, Mark Harrison, and Steve Sturdy, eds., *Medicine and Modern Warfare* (Amsterdam: Rodopi, 1999), 185–204.

23. For more on the relationship between Mack and Lathrop, see Jane Addams, *My Friend, Julia Lathrop* (New York: Macmillan, 1935), 136. For more on female reformers in Chicago, including Lathrop, see Maureen Flanagan, *Seeing with Their Hearts: Chicago Women and the Vision of the Good City, 1871–1933* (Princeton, NJ: Princeton University Press, 2002).

24. Harry Barnard, *The Forging of an American Jew: The Life and Times of Judge Julian W. Mack* (New York: Herzl Press, 1974), 209.

25. The most comprehensive account of the history of the War Risk Insurance Act is K. Walter Hickel, "Entitling Citizens: World War I, Progressivism, and the Origins of the American Welfare State, 1917–1928" (PhD diss., Columbia University, 1999). See also his "War, Region, and Social Welfare: Federal Aid to Servicemen's Dependents in the South, 1917–1921," *Journal of American History* 87 (March 2001): 1362–91.

26. The medical specialty of physical medicine, or, as it is sometimes called, physiatry, did not gain recognition from the American Medical Association until 1949. For more, see Rosemary A. Stevens, *American Medicine and the Public Interest*, 2nd ed. (Berkeley: University of California Press, 1998).

27. The one exception, of course, would be the National Homes for Civil War veterans. Patrick J. Kelly, *Creating a National Home: Building the Veterans' Welfare State, 1860–1900* (Cambridge, MA: Harvard University, 1997). The purpose of these homes was to provide food and shelter for poor and destitute veterans. Veterans who took up residence in the National Homes were not expected to overcome their disabilities and state of poverty so that they could become self-sufficient land- and homeowners again.

28. The Office of the Surgeon General, *The Medical Department of the United States Army in the World War* 1 (Washington, DC: GPO, 1923): esp. 474–85. Hereafter cited throughout the notes as *MDWW*.

29. The literature on psychology, psychiatry, shellshock, and World War I is quite extensive. See, for example, Paul Frederick Lerner, *Hysterical Men: War, Psychiatry, and the Politics of Trauma in Germany, 1890–1930* (Ithaca, NY: Cornell University Press, 2003); Mark S. Micale, *Hysterical Men: The Hidden History of Male Nervous Illness* (Cambridge, MA: Harvard University Press, 2008); and Peter Leese, *Shell Shock: Traumatic Neurosis and the British Soldiers of the First World War* (Hampshire and New York: Palgrave Macmillan, 2002). See also *Journal of Contemporary History* 35, no. 1 (2000), a special issue edited by Jay Winter that provides a cross-cultural analysis of shellshock and the war. For the rise of psychiatry in the United States, see Elizabeth Lunbeck, *The Psychiatric Persuasion: Knowledge, Gender, and Power in Modern America* (Princeton, NJ: Princeton University Press, 1994). For the role of psychologists in World War I, see Leila Zenderland, *Measuring Minds: Henry Herbert Goddard and the Origins of American Intelligence Testing* (Cambridge: Cambridge University Press, 1998), and Daniel J. Kevles, "Testing the Army's Intelligence: Psychologists and the Military in World War I," *Journal of American History* 55 (1968): 565–81.

30. Very little has been written about the treatment of war-related sensory disabilities. For the rehabilitation of blinded soldiers in Britain, see Julie Anderson and Neil Pemberton, "Walking Alone: Aiding the War and the Civilian Blind in the Inter-war Period," *European Review of History* 14, no. 4 (December 2007): 459–79.

31. For more, see Carol R. Byerly, *Good Tuberculosis Men: The Army Medical Department's Struggle with Tuberculosis* (Washington, DC: Borden Institute, forthcoming.)

32. For this line of argument, see Douglas McMurtrie, *The Disabled Soldier* (New York: Macmillan, 1919), 19.

33. Medical ideas, social reform, cultural values, professional authority, and policymaking all coalesced to create the system of rehabilitation for disabled soldiers in the United States. Contrary to certain scholars, I do not

see rehabilitation primarily as a medical specialty. For examples of rehabilitation as a medical specialty, see Glenn Gritzer and Arnold Arluke, *The
Making of Rehabilitation: A Political Economy of Medical Specialization, 1890–
1980* (Berkeley: University of California Press, 1985). See also Paul Starr,
The Social Transformation of American Medicine (New York: Basic Books,
1982), Rosemary Stevens, *American Medicine and the Public Interest*, 2nd ed.
(Berkeley: University of California Press, 1998), and George Weisz, *Divide
and Conquer: A Comparative History of Medical Specialization* (Oxford: Oxford
University Press, 2006).

34. See *MDWW* 1: 482.

35. I am mindful of the insistence among disability scholars that every story
must include the voice of the disabled themselves, a slogan that appears in
the title of James I. Charlton's book, *Nothing about Us without Us: Disability Oppression and Empowerment* (Berkeley: University of California Press,
2000). Scholars of the social history of medicine have rightly issued similar
directives, beginning with Roy Porter's 1985 call for more patient-centered
accounts; see "The Patient's View: Doing Medical History from Below,"
Theory and Society 14 (1985): 167–74. The sentiment in favor of "below"
history is expressed widely in the field of disability history. See, for example, Susan Burch and Ian Sutherland, "Who's Not yet Here? American
Disability History," *Radical History Review* 94 (2006): 127. The history of
disabled veterans is often seen as a separate subset of the field. For the
best historical treatment of US veterans, see David A. Gerber, ed., *Disabled
Veterans in History* (Ann Arbor: University of Michigan Press, 2000). I found
soldier-patient testimony in Army Inspectors General Reports held at the
National Archives and Records Administration, College Park, Maryland.
When a patient or someone representing a patient made a complaint about
medical care, the inspectors general would be alerted, and an investigation
would take place that often included dozens of patients' testimony. These
reports also often include patient medical records. For more on the role
and history of the inspectors general, see Joseph W. A. Whitehorne, *The
Inspectors General of the United States Army, 1903–1939* (Washington, DC:
Office of the Inspector General and Center of Military History, United
States Army, 1998).

36. Some will regard this book as a case study in medicalization. I do not deny
the fact that veteran welfare became medicalized during the First World
War, affecting how disabled soldiers would be understood for the remainder of the twentieth century. I would, however, insist that the medicalizing
process be understood from a multitude of perspectives, not simply as a
top-down process whereby physicians, eager for financial gain and institutional control, reduce patients to biological entities without considering
the social, political, personal, and economic ramifications of disease and
disability. Disability and illness are shared phenomena that cut across
class, gender, and race. Many of the physicians featured in this book were

disabled themselves, suffering from the same physical diseases and deformities as their patients. The treatments, prostheses, and braces that certain
orthopedic surgeons prescribed for others, they also used themselves. In
other words, the physicians (often assumed to be at the "top" in bottom-up
histories) provide one perspective on what it meant to be disabled in the
early twentieth century.

Early theorists understood medicalization to be beneficial only for
power-seeking, money-hungry doctors but deleterious for patients and society. See, for example, Ivan Illich, *Medical Nemesis* (London: Marion Boyars,
1976); Irving K. Zola, "Medicine as an Institution of Social Control," *Sociological Review* 20 (1972): 487–504; and Thomas Szasz, *The Manufacture of
Madness* (New York: Harper Row, 1970). Renée Fox proved to be an exception at the time. See Fox, "The Medicalization and Demedicalization of
American Society," *Daedalus* 106 (Winter 1977): 9–22. More recent examples of this Foucauldian-inspired argument can be found among disability
historians who insist that the institution of rehabilitation is driven by the
medical model. Brad Byrom makes this argument in "A Pupil and a Patient:
Hospital-Schools in Progressive America," in Paul K. Longmore and Lauri
Umansky, eds., *The New Disability History: American Perspectives* (New York:
New York University Press, 2001), 133–56. For the best descriptions of the
"medical model," see Paul K. Longmore and Lauri Umansky, "Disability
History: From the Margins to the Mainstream," in *The New Disability His-
tory*, 1–29, and Catherine J. Kudlick, "Why We Need Another 'Other,'"
American Historical Review 108 (June 2003): 763–93.

I find Peter Conrad's recent work on medicalization much more astute
and persuasive than the theories espoused over thirty years ago. According
to Conrad, direct physician involvement is not a necessary condition to
medicalization. Moreover, Conrad maintains that some forms of medicalization are more humanistic than other forms of social intervention, such
as incarceration. See Conrad, *Medicalization of Society: On the Transformation of Human Conditions into Treatable Disorders* (Baltimore: Johns Hopkins
University Press, 2007), and Conrad and Joseph Schneider, *Deviance and
Medicalization: From Badness to Sickness*, 2nd ed. (Philadelphia: Temple University Press, 1992). Universal health care, I would maintain, is one of the
best examples of how medicalization can be beneficial.

37. Numbers and percentages extrapolated from *MDWW* 1: 480–81.
38. The best historical treatment of veteran welfare policy after the First World
War is Stephen R Ortiz, *Beyond the Bonus March and GI Bill: How Veteran
Politics Shaped the New Deal Era* (New York: New York University Press,
2010).
39. David Silver, Washington, DC, to Major Joel E. Goldthwait, AEF [American
Expeditionary Forces], France, 11 December 1917, National Archives and
Records Administration, College Park, Maryland, Record Group 112, box
309, file 442.3.

40. http://www.frontrunnermagazine.com/2009/10/08/rest-when-youre-dead-the-story-of-garth-stewart/.

CHAPTER ONE

1. For accounts from the *Los Angeles Times*, see "Wilson Will Review Grand Army Veterans," September 13, 1915, 11; "G.A.R. Hosts in Washington," September 27, 1915, 13; "Review of G.A.R. Thrills Capital," 30 September 1915, 5; "Washington Hails Arriving Veterans," 27 September 1915, 11; "Veterans Tread Old-Time Trails," September 30, 1915, 15. For *New York Times* accounts of the parade, see "Wilson Interprets Ideals to Veterans," September 29, 1915, 4; "Review of G.A.R. Thrills Capital," September 30, 1915, 5; "Washington Hails Arriving Veterans," September 27, 1915, 11; "To Repeat Grand Review: 40,000 Veterans of Civil War Will March on Fiftieth Anniversary," August 16, 1915, 10; and Edward Marshall, "Thinning Blue Line Fifty Years after Peace," September 26, 1915, SM16.

2. For the report of Wilson's tears, see "Grand Army of the Republic Grand Parade," *Washington Herald,* September 30, 1915. http://suvcw.org/banner/1915parade.htm. See also "G.A.R. Veterans Cheer Wilson," *New York Times,* September 29, 1915, 13, and "Review of G.A.R. Thrills Capital," *New York Times,* 30 September 1915, 5.

3. Kendrick A. Clements, *The Presidency of Woodrow Wilson* (Lawrence: University Press of Kansas, 1992). For general overviews of the United States and the First World War, see David M. Kennedy, *Over Here: The First World War and American Society* (New York: Oxford University Press, 1980), and Jennifer D. Keene, *Doughboys, the Great War, and the Remaking of America* (Baltimore: Johns Hopkins University Press, 2001).

4. Clements, *Woodrow Wilson,* 128.

5. For "a mother lode of nostalgia," see David Blight, *Race and Reunion: The Civil War in American Memory* (Cambridge, MA: Belknap Press of Harvard University Press, 2001), 4. For more on the cultural salience of Civil War commemoration and memories, see John Pettegrew, "'The Soldier's Faith': Turn-of-the-Century Memory of the Civil War and the Emergence of Modern American Nationalism," *Journal of Contemporary History* 31, no. 1 (1996): 49–73.

6. William M. Sloane, "Pensions and Socialism," *Century Illustrated Magazine* 42 (June 1891): 183.

7. For these data, see Theda Skocpol, *Protecting Soldiers and Mothers: The Political Origins of Social Policy in the United States* (Cambridge, MA: Harvard University Press, 1992), 109–10, and William H. Glasson, *Federal Military Pensions in the United States* (New York: Oxford University Press, 1918), esp. chap. 3 and the statistical tables found on 273.

8. Confederates would not receive pensions under the General Law until 1958.

See Jeffrey E. Vogel, "Redefining Reconciliation: Confederate Veterans and the Southern Responses to Federal Civil War Pensions," *Civil War History* 51, no. 1 (2005): 89.

9. Larry M. Logue and Peter Blanck, "'Benefit of the Doubt': African-American Civil War Veterans and Pensions," *Journal of Interdisciplinary History* 38, no. 3 (2008): 377–99. Theda Skocpol argues otherwise. According to Skocpol, African Americans made up about 9 to 10 percent of the Union forces. "Although there is no systematic evidence about how black Union veterans fared in the pension application process compared to whites," writes Skocpol, "hints from the historical record suggest that free blacks with stable residential histories in the North probably did as well as their white socioeconomic counterparts, while black veterans and survivors from the ranks of freed slaves may often have lacked the documents they needed to establish claims for pensions" (*Protecting Soldiers and Mothers*, 138).

10. Burton J. Hendrick, "Pork-Barrel Pensions," *World's Work* 30 (October 1915): 719.

11. Sloane, "Pensions and Socialism," 184.

12. Skocpol, *Protecting Soldiers and Mothers*, 106.

13. John P. Resch, "Politics and Public Culture: The Revolutionary War Pension Act of 1818," *Journal of the Early Republic* 8 (Summer 1988): 141.

14. *Journal of the House of Representatives*, 15th Cong., 1st sess., 512; *Annals of Congress*, 15th Cong., 1st sess., 497–99, as quoted in Resch, "Politics and Public Culture," 143.

15. Resch, "Politics and Public Culture," 141.

16. Ibid., 148.

17. *Annals of Congress*, 15th Cong., 1st sess., 141, as quoted in Resch, "Politics and Public Culture," 148–49.

18. For a description of the rush of applications, see President's Commission on Veterans' Pensions, *The Historical Development of the Veterans' Benefit in the United States* (Washington, DC: GPO, 1956), 8.

19. Resch, "Politics and Public Culture," 157.

20. For "humiliating process," see *Annals of Congress*, 15th Cong., 1st sess., 497–99, as quoted in Resch, "Politics and Public Culture," 143.

21. One of best historical treatments of how the General Law was implemented is Larry M. Logue and Peter Blanck, *Race, Ethnicity, and Disability: Veterans and Benefits in Post–Civil War America* (Cambridge: Cambridge University Press, 2010).

22. By 1891, the Pension Bureau employed 2,000 citizens to work as clerical workers and special examiners. See P. D. Blanck and M. Millender, "Before Disability Civil Rights: Civil War Pensions and the Politics of Disability in America," *Alabama Law Review* 52 (2000): 11.

23. Patrick J. Kelly, *Creating a National Home: Building the Veterans' Welfare State, 1860–1900* (Cambridge, MA: Harvard University, 1997), 10.

24. Kelly, *Creating a National Home,* 198.
25. Patrick J. Kelly, "The Election of 1896 and the Restructuring of the Civil War Memory," *Civil War History* 49 (2003): 254–80.
26. Skocpol makes this argument, as do Blanck and Millender.
27. R. B. Rosenburg, "'Empty Sleeves and Wooden Pegs': Disabled Confederate Veterans in Image and Reality," in David A. Gerber, ed., *Disabled Veterans in History* (Ann Arbor: University of Michigan Press, 2000), 218.
28. Blanck and Millender, "Before Disability Civil Rights," 33.
29. Rosenburg, "'Empty Sleeves and Wooden Pegs': Disabled Confederate Veterans in Image and Reality," 221.
30. For a good survey of Gilded Age political history, see Mark W. Summers, *Party Games: Getting, Keeping, and Using Power in Gilded Age Politics* (Chapel Hill: University of North Carolina Press, 2004). See also C. W. Calhoun, "Reimagining the 'Lost Men' of the Gilded Age: Perspectives on the Late Nineteenth Century Presidents," *Journal of the Gilded Age and Progressive Era* 1, no. 3 (July 2002): 225–57.
31. Blanck and Millender, "Before Disability Civil Rights," 8.
32. Figures taken from Skocpol, *Protecting Soldiers and Mothers,* 116.
33. Curtis, "The Mugwump and the Bourbon," *Harper's Weekly* 11 (1885): 755.
34. Skocpol, *Protecting Soldiers and Mothers,* 127.
35. Donald L. McMurry, "The Political Significance of the Pension Question, 1885–1897," *Mississippi Valley Historical Review* 9 (June 1922): 30.
36. Skocpol, *Protecting Soldiers and Mothers,* 127.
37. Ibid., 128.
38. McMurry, "The Political Significance of the Pension Question, 1885–1897," 29 and 35.
39. Sloane, "Pensions and Socialism," 184.
40. Ibid., 185 and 183.
41. Ibid., 184.
42. Ibid., 186.
43. Skocpol, *Protecting Soldiers and Mothers,* 129.
44. Carole Haber, *Beyond Sixty-Five: The Dilemma of Old Age in America's Past* (New York: Cambridge University Press, 1983), 111–13.
45. Skocpol, *Protecting Soldiers and Mothers,* 132.
46. For a wonderfully engaging biography of Page, see John Milton Cooper, Jr., *Walter Hines Page: The Southerner as American, 1855–1918* (Chapel Hill: University of North Carolina Press, 1977).
47. For an overview, including circulation numbers of these magazines, see Theodore Bernard Peterson, *Magazines in the Twentieth Century* (Urbana: University of Illinois Press, 1964).
48. "A Degrading Conception of Pensions," *World's Work* 7 (November 1903–4): 4508.
49. Cooper, *Walter Hines Page,* xxiv.

50. The literature on the history of the concept of the New South is vast. A classic is C. Vann Woodward, *Origins of the New South* (Baton Rouge: Louisiana State University Press, 1951). For a more recent treatment, see Edward Ayers, *The Promise of the New South: Life after Reconstruction* (Oxford: Oxford University Press, 2007).
51. "A Degrading Conception of Pensions," 4508.
52. Glasson, *Federal Military Pensions*, 267.
53. Virginia E. McCormick, "The Talented Sherwoods: Poets and Politicians," *Northwest Ohio Quarterly* 52, no. 3 (1980): 244–53.
54. Glasson, *Federal Military Pensions*, 254–59.
55. "Another Pension Bill," *World's Work* 28 (May 1914): 12.
56. Figures taken from Glasson, *Federal Military Pensions*, 259.
57. "Democrats and Pensions," *World's Work* 29 (February 1915): 372.
58. "The Progressive Programme," *World's Work* 24 (September 1912): 489–91, 490.
59. For more about the masculinity crisis, see Gail Bederman, *Manliness and Civilization: A Cultural History of Gender and Race in the United States, 1880–1917* (Chicago: University of Chicago Press, 1995); Kristin Hoganson, *Fighting for American Manhood: How Gender Politics Provoked the Spanish-American and Philippine-American Wars* (New Haven: Yale University Press, 1998); John F. Kasson, *Houdini, Tarzan, and the Perfect Man: The White Male Body and the Challenge of Modernity in America* (New York: Hill and Wang, 2001); and T. J. Jackson Lears, *No Place of Grace: Antimodernism and the Transformation of American Culture, 1880–1920*, 2nd ed. (Chicago: University of Chicago Press, 1994), and *Rebirth of a Nation: The Making of Modern America, 1877–1920* (New York: HarperCollins, 2009).
60. Hoganson, *Fighting for American Manhood*, 10.
61. Ibid., 23.
62. Henry Richard Gibson, *Liberality, Not Parsimony: This Is Our True Pension Policy* (Washington, DC: GPO, 1900).
63. Hendrick, "Pork-Barrel Pensions," 102.
64. Ibid.
65. For numbers and figures see Skocpol, *Protecting Soldiers and Mothers*, 109–10, and Glasson, *Federal Military Pensions*, esp. chap. 3 and the statistical tables found on 273.
66. Hendrick, "Pork-Barrel Pensions," 101.
67. Mildred Maroney, "Veterans' Benefits," in Lewis Meriam and Karl T. Schlotterbeck, eds., *The Cost and Financing of Social Security* (Washington, DC: Brookings Institution, 1950), 103.
68. M. Lincoln Schuster, "The Passing of Pension Plunder," *World's Work* 34 (October 1917): 689–92, 689.
69. For the most comprehensive account of the Section on Compensation for Soldiers and Sailors, see Karl Walter Hickel, "Entitling Citizens: World

War I, Progressivism, and the Origins of the American Welfare State, 1917–1928" (PhD diss., Columbia University, 1999).

70. Julia C. Lathrop, "The Military and Naval Insurance Act," *Nation* 106 (7 February 1918): 157–58. Lathrop also testified before Congress in support of passing the War Risk Insurance Act of 1917. See "Statement of Miss Julia Lathrop," in US Congress, House of Representatives Committee on Interstate and Foreign Commerce, *To Amend the Bureau of Insurance Act so as to Insure the Men in the Army and Navy*, 65th Cong., 1st sess., 17 August 1917, 127–44.

71. For more about Mack's career and life, see Harry Barnard, *The Forging of an American Jew: The Life and Times of Judge Julian W. Mack* (New York: Herzl Press, 1974).

72. K. Walter Hickel, "War, Region, and Social Welfare: Federal Aid to Servicemen's Dependents in the South, 1917–1921," *Journal of American History* 87, no. 4 (March 2001): 1366.

73. Samuel McCune Lindsay, "Purpose and Scope of War Risk Insurance," *Annals of the American Academy of Political and Social Science* 79 (September 1918): 52–68, 60.

74. Rosemary Stevens, "Can the Government Govern? Lessons from the Formation of the Veterans Administration," *Journal of Health Politics, Policy, and Law* 16 (Summer 1991): 285.

75. During his first term (1912–16), Wilson signed into law pension amendments that provided more liberal provisions to war widows, especially those who had married after 1890 and whose husbands had not died from causes originating from military service. W. L. Stoddard of the *New Republic* expressed doubt about Wilson's commitment to pension reform, especially as he neared reelection. In the spring of 1915, a bill that promised to provide a greater monthly payment to Civil War and Spanish-American War widows was working its way through Congress. Forecasting its passage in Congress, Stoddard wrote that Wilson would face a "difficult dilemma" when it arrived on his desk to be signed into law. "The weight of precedent" argued Stoddard, "will bend him toward obeying the will of the pensioners; the weight of the pitifully small and unorganized pension reform movement will bend him in the opposite direction." Believing that Progressive reformers were no match for the "powerful pension interests," he concluded that in order to veto the bill, Wilson would have had to be "more than human." True to Stoddard's prediction, Wilson signed the War Risk Insurance Act of September 8, 1916, into law. W. L. Stoddard, "More Federal Pensions," *New Republic* 1 (March 6, 1915): 126. For more on Wilson's own history with the Civil War, see A. Gaughan, "Woodrow Wilson and the Legacy of the Civil War," *Civil War History* 43, no. 3 (1997): 225–42.

76. Theodore Price, "The Great Insurance Adventure," *Outlook* 120 (1919): 20.

77. "Books in Brief," *Nation* 109 (12 July 1919): 49–50.

78. Schuster, "The Passing of Pension Plunder," 692.
79. Samuel McCune Lindsay, "Soldiers' and Sailors' Insurance," *Proceedings of the American Philosophical Society* 57 (1918): 632.
80. Hickel also makes this point in both "Entitling Citizens," and "War, Region, and Social Welfare."
81. See reprint of the Act in Glasson, *Federal Military Pensions*, 287–95. The provision for mandatory medical care can be found under section 302; see Glasson, 291–92.
82. Jeffry Kostic, "War Risk Insurance," in *The United States in the First World War: An Encyclopedia*, ed. Anne Cipriano Venzon (New York: Garland Publishing, 1995), 776. See also Jennifer D. Keene, *World War I* (Westport, CT: Greenwood Press, 2006), 45–46.
83. Hickel, "War, Region, and Social Welfare," 1373.
84. Kostic, "War Risk Insurance," 776. Worried that only a small percentage of servicemen would take advantage of this opportunity, the War Risk Insurance Bureau embarked on a massive insurance drive, posting advertisements in movie houses and battleships, as well as sending agents into the trenches to get soldiers who were already overseas enrolled. See Theodore H. Price and Richard Spillane, "The Great Insurance Adventure," *Outlook* 121 (1919): 20, and P. H. Douglas, "The War Risk Insurance Act," *Journal of Political Economy* (1918): 479.
85. Contemporary analysts who supported the WRIA repeatedly made the point that it was far more fiscally conservative than a pension system. For some examples of supporters using this line of argumentation, see T. B. Love, "The Social Significance of War Risk Insurance," *Annals of the American Academy of Political and Social Science* 79 (1918): 46–51; S. H. Wolfe, "Eight Months of War Risk Insurance Work," *Annals of the American Academy of Political and Social Science* 79 (1918): 68–79; and C. Harold Waterbury, "Modern Forms of Insurance Protection," *Nation* 106 (7 February 1918): 164–65.
86. Hickel, "Entitling Citizens," 139.
87. Hickel argues that with the WRIA, there was a marriage, of sorts, between paternalist and maternalist social reformers. He also points out that a lot was riding on the WRIA for the male academic reformers who promoted paternalistic policies because they had only achieved a modicum of success compared with female, maternalist reformers. See Hickel, "Entitling Citizens," 121–22, and "War, Region, and Social Welfare," 1367. For a contemporary maternalist's viewpoint of the WRIA, see Lathrop, "The Military and Naval Insurance Act." For more on maternalism and the role that women played in welfare reform, see Sonya Michel, "The Limits of Maternalism: Policies toward American Wage-Earning Mothers during the Progressive Era," in Seth Koven and Sonya Michel, eds., *Mothers of a New World: Maternalist Politics and the Origins of Welfare States* (New York: Routledge, 1993), 277–320, and Seth Koven and Sonya Michel, "Womanly Duties: Maternalist

Politics and the Origins of Welfare States in France, Germany, Great Britain, and the United States, 1880–1920," *American Historical Review* 95 (October 1990): 1076–1108.

88. Hickel, "War, Region, and Social Welfare," 1374.

89. Lindsay, "Purpose and Scope," 62.

90. See Michael Willrich's illuminating article about how Progressive-Era insistence on male wage-earning resulted in many punitive policies and discrimination targeted at unemployed men; "Home Slackers: Men, the State, and Welfare in Modern America," *Journal of American History* 87 (September 2000): 460–89.

91. "Meeting of Insurance Committee Held in Office of the Secretary of the Treasury July 25, 1917," minutes, p. 11, file "War Risk Insurance and Treasury Reports," box 562, Subject Files, Papers of William Gibbs McAdoo, Library of Congress, as quoted in Hickel, "Entitling Citizens," 118.

92. See section 302 of the act in Glasson, *Federal Military Pensions*, 291–92.

93. Thomas Gregory, "Restoring Crippled Soldiers to a Useful Life," *World's Work* 36 (July 1918): 428.

94. John Culbert Faries, *The Economic Consequences of Physical Disability: A Case Study of Civilian Cripples in New York City* (New York City: Red Cross Institute for Crippled and Disabled Men, 1918).

95. Douglas, "The War Risk Insurance Act," 475.

96. Wolfe, "Eight Months of War Risk Insurance Work," 75.

97. This confirms Daniel T. Rodgers' argument that Progressivism was a transnational phenomenon. See Rodgers, *Atlantic Crossings: Social Politics in a Progressive Age* (Cambridge, MA: Belknap Press of Harvard University Press, 1998).

98. Douglas, "The War Risk Insurance Act," 475.

CHAPTER TWO

1. Michael J. Lansing, " 'Salvaging the Man Power of America: Conservation, Manhood, and Disabled Veterans During World War I," *Environmental History* 14 (2009): 32–57, 34. The rehabilitation movement should be seen as an important component of the increasing Progressive Era concern about labor power, a story best illustrated in Anson Rabinbach, *The Human Motor: Energy, Fatigue, and the Origins of Modernity* (Berkeley: University of California Press, 1992).

2. This number is based on American Orthopedic Association membership numbers. See Thornton Brown, *The American Orthopaedic Association: A Centennial History* (Park Ridge, IL: American Orthopaedic Association, 1987), 9. For medical specialization in the early to mid nineteenth century in the United States, see Rosemary A. Stevens, *American Medicine and the Public Interest*, 2nd (Berkeley: University of California Press, 1988), and George

Weisz, *Divide and Conquer: A Comparative History of Medical Specialization* (Oxford: Oxford University Press, 2006), esp. chap. 4.

3. The records of the Philadelphia Orthopaedic Hospital and Infirmary for Nervous Diseases are held at Historical Medical Library of the College of Physicians of Philadelphia. For a ledger of patients admitted to the hospital, including their age, condition, and procedure performed, see MSS 6, box 14.

4. For more on the history of children and rehabilitation, see Walton O. Schalick, "Children, Disability, and Rehabilitation in History," *Pediatric Rehabilitation* 4 (2001): 91–95. See also Douglas C. McMurtrie, "The Care of Crippled Children in the United States," *American Journal of Orthopedic Surgery* 9 (1912): 527–56.

5. For histories of rehabilitation in Great Britain, see Roger Cooter, *Surgery and Society in Peace and War: Orthopaedics and the Organization of Modern Medicine, 1880–1948* (London: Macmillan Press, 1993); Seth Koven, "Remembering and Dismemberment: Crippled Children, Wounded Soldiers, and the Great War in Great Britain," *American Historical Review* 99 (1994): 1167–1202; Jeffrey Reznick, *Healing the Nation: Soldiers and the Culture of Caregiving in Britain during the Great War* (Manchester: Manchester University Press, 2004); Deborah Cohen, *The War Come Home: Disabled Veterans in Britain and Germany, 1914–1939* (Berkeley: University of California Press, 2001); Ana Carden-Coyne, *Reconstructing the Body: Classicism, Modernism, and the First World War* (Oxford: Oxford University Press, 2009); and Joanne Bourke, *Dismembering the Male: Men's Bodies, Britain, and the Great War* (London: Reaktion Books, 1996). Other works on World War I disabled veterans in Europe include Thomas Schlich, "The Perfect Machine: Lorenz Böhler's Rationalized Fracture Treatment in WWI," *Isis* 100 (2009): 758–91; Roxanne Panchasi, "Reconstructions: Prosthetics and the Rehabilitation of the Male Body in World War I France," *Differences: A Journal of Feminist Cultural Studies* 7 (1995): 109–40; and Heather R. Perry, "Recycling the Disabled: Army, Medicine, and Society in World War I Germany" (PhD diss., Indiana University, 2005). For a post-structuralist critique of the rise of rehabilitation, with an emphasis on France, see Henri-Jacques Striker, *A History of Disability*, trans. William Sayers (Ann Arbor: University of Michigan Press, 1999), esp. chap. 6. Sociologists Glen Gritzer and Arnold Arluke provide a brief history of how rehabilitation became a crucial part of the health care delivery system in the United States beginning with the First World War; see *The Making of Rehabilitation: A Political Economy of Medical Specialization, 1890–1980* (Berkeley: University of California Press, 1985). Ana Carden-Coyne researches US rehabilitation efforts from the soldiers' perspective. See, for example, "Ungrateful Bodies: Rehabilitation, Resistance, and Disabled American Veterans of the First World War," *European Review of History* 14, no. 4 (2007): 543–65.

6. Membership rosters can be found in the yearly publications of the *Transactions of the American Orthopaedic Association* from 1889 to 1902, followed by the *American Journal of Orthopedic Surgery* from 1903 to 1918.
7. For the importance of trans-Atlantic relationships to Progressive Era movements, see Daniel T. Rodgers, *Atlantic Crossings: Social Politics in a Progressive Age* (Cambridge, MA: Belknap Press of Harvard University Press, 1998).
8. *MDWW* 11 (Washington, DC: GPO, 1927): 550. The exact wording for the creation of the Division of Orthopedic Surgery reads as follows: "for the purpose of providing disabled soldiers with 'orthopedic reconstruction,'" as well as an "industrial reeducation . . . to fit them for return to civil life."
9. For the history of pre–World War I military medical care, see Mary C. Gillett, *The Army Medical Department, 1865–1917* (Washington, DC: Center of Military History, United States Army, 1995).
10. Gritzer and Arluke see the army's creation of the Division of Orthopedics as crucial to the professionalization of orthopedic surgery in the United States. As I outline below, I find the profession's prewar involvement with the rehabilitation of crippled children just as, if not more, important. See Glenn Gritzer and Arnold Arluke, *The Making of Rehabilitation,* 40ff.
11. My use of the word "crippled" is in keeping with the language of my historical actors. The story of Progressives penalizing the disabled who physically displayed their maimed bodies in order to gain money begging on city streets is best exemplified in Susan M. Schweik, *The Ugly Laws: Disability in Public* (New York: New York University, 2009). Between 1880 and 1920, more than a handful of cities across the United States passed ordinances, fining any "unsightly or disgusting" individual who "exposed" him- or herself to the public; see *Ugly Laws,* 1–3.
12. Gwilym G. Davis "President's Address," *American Journal of Orthopedic Surgery* 12 (July 1914): 1.
13. For a comprehensive history of welfare reform in the United States, spanning the nineteenth and twentieth centuries, see Michael Katz, *In the Shadow of the Poorhouse: A Social History of Welfare in America* (New York: Basic Books, 1986).
14. Emphasis on the work ethic persisted, as Daniel T. Rodgers shows, well into the Progressive-Era and manifested itself in the industrial workplace; see *The Work Ethic in Industrial America, 1850–1920* (Chicago: University of Chicago Press, 1978).
15. For "mental and moral training," see Davis, "President's Address," 4.
16. For "social guardians," see Davis, "President's Address," 3.
17. The literature in the history of medicine and surgery that covers the discovery of the germ theory of disease and antiseptics is quite extensive. For germ theory, some good examples include: David Barnes, *The Great Stink of Paris* (Baltimore: Johns Hopkins University Press, 2006); Gerald Geison, *The Private Science of Louis Pasteur* (Princeton: Princeton University Press, 1995); Nancy Tomes, *The Gospel of Germs* (Cambridge, MA: Harvard Uni-

versity Press, 1998). English surgeon Joseph Lister published his seminal paper on antisepsis in an 1867 issue of *The Lancet*. On the whole, American surgeons were slow to adopt antisepsis. For more, see Christoph Mörgeli, *The Surgeon's Stage* (Basel: Roche, 1999), 203–32; Thomas P. Gariepy, "The Introduction and Acceptance of Listerian Antisepsis in the United States," *Journal of the History of Medicine and Allied Sciences* 49 (1994): 167–206; and Gert H. Brieger, "A Portrait of Surgery: Surgery in America 1875–1889," *Surgical Clinics of North America* 67 (December 1987): 1181–1215. See also Christopher Lawrence, ed., *Medical Theory, Surgical Practice* (London: Routledge, 1992), and Martin Pernick, *A Calculus of Suffering* (New York: Columbia University Press, 1985).

18. New operative techniques in orthopedic surgery were beginning to take hold at the turn of the twentieth century. Russell Hibbs of the New York Orthopedic Hospital began performing spinal fusions for Pott's disease patients in 1911; see R. A. Hibbs, "An Operation for Progressive Spinal Deformities," *New York Medical Journal* 93 (1911): 1013–16. Nevertheless, US orthopedic surgeons, unlike some of their foreign counterparts, emphasized mechanical treatment over operative work. Surgeons like Hibbs, who specialized in surgical operations rather than noninvasive treatments, were not considered to be "true" orthopedic surgeons. Indeed, the AOA denied Hibbs membership until well after the First World War. Experimental joint replacements began as early as the 1890s, but would not become common practice to orthopedic surgery until the 1920s. One can trace this history using primary sources such as the multiple editions of Edward Bradford and Robert Lovett's *Treatise on Orthopedic Surgery*, 1st ed. (New York: William Wood & Co., 1890). The fifth edition was published in 1915. For secondary source reading on the subject, see Julie Anderson, Francis Neary, and John V. Pickstone, *Surgeons, Manufacturers, and Patients* (Houndmills, Basingstoke: Palgrave Macmillan, 2007).

19. I see medical welfarism as similar to welfare capitalism in the sense that both assumed that a healthy worker was a happy worker who would not engage in labor strikes and revolts. For more on the scientific charity movement of the Progressive Era, see Katz, *In the Shadow of the Poorhouse*, esp. chap. 3; and Emily K. Abel, "Valuing Care: Turn-of-the-Century Conflicts between Charity Workers and Women Clients," *Journal of Women's History* 10 (Autumn 1998): 32–52.

20. This number represents white infants only; see Michael R. Haines, "Fertility and Mortality, by Race: 1800–2000," in Susan B. Carter, ed., *Historical Statistics of the United States*, vol. 1 (Cambridge: Cambridge University Press, 2006), table Ab1-10.

21. Peter C. English, "Not Miniature Men and Women," in Loretta M. Kopelman and John C. Moskop, eds., *Children and Health Care: Moral and Social Issues* (Dordrecht: Kluwer, 1989), 252.

22. For overviews on the history of child health and health care, see Janet

Golden, Richard A. Meckel, and Heather Munro Prescott, *Children and Youth in Sickness and in Health* (Westport, CT: Greenwood Press, 2004); Alexandra Minna Stern and Howard Markel, eds., *Formative Years* (Ann Arbor: University of Michigan Press, 2002); and Roger Cooter, ed., *In the Name of the Child* (London: Routledge, 1992). For a history of the profession of pediatrics, see Sydney Halpern, *American Pediatrics* (Berkeley and Los Angeles: University of California Press, 1988).

23. "Dr. James Knight's Death," *New York Times*, October 25, 1887.

24. Elliot Brackett, "Meeting of the Boston Orthopedic Club, Oct. 6, 1917," *American Journal of Orthopedic Surgery* 15 (1917): 843.

25. James Knight, *Orthopaedia, or a Practical Treatise on the Aberrations of the Human Form* (New York: G. P. Putnam's Sons, 1874).

26. Orthopedic surgeons did use cerebral paralysis as a diagnostic category in the late nineteenth century. For more on the accepted etiology of the condition, see Bradford and Lovett, *Treatise*.

27. At the time the exact etiology of rickets remained unknown. Physicians at the time did correlate deficient diet and unhealthy environments to disease causation and often prescribed heliotherapy and cod-liver oil in order to cure the disease. That rickets resulted from a vitamin deficiency, however, would not be discovered until the second decade of the twentieth century when physicians-experimenters who were interested in so-called chemical hygiene would isolate the cause. For a more comprehensive discussion of the history of rickets, see Rima Apple, *Vitamania: Vitamins in American Culture* (New Brunswick, NJ: Rutgers University Press, 1996); as well as Sally D. Romano, "The Dark Side of the Sun: Skin Cancer, Sunscreen, and Risk in Twentieth-Century America" (PhD diss., Yale University, 2006), esp. chap. 1.

28. Pott's disease was named after Percival Pott (1714–88), who, well before Robert Koch's actual discovery of the *tubercle bacillus* in 1882, first identified the spinal affliction as one that originated from an unspecified disease process and not from acute trauma. While much scholarship in the history of medicine as been devoted to respiratory tuberculosis (TB), comparatively little has been written about tuberculosis of the bone. The delivery of health care at the institutional level was similar for both groups of patients, that is, both were treated in sanatoriums and open-air hospitals. But the actual day-to-day care differed dramatically in that tuberculosis of the bone was treated primarily through orthopedic means, whereas respiratory TB was not. Some sample works on the history of respiratory tuberculosis in the United States include: Carol R. Byerly, *Good Tuberculosis Men: The Army Medical Department's Struggle with Tuberculosis* (Washington, DC: Borden Institute, forthcoming); Sheila Rothman, *Living in the Shadow of Death: Tuberculosis and the Social Experience of Illness in American History* (New York: Basic Books, 1994); Michael E. Teller, *The Tuberculosis Movement: A Public Health Campaign in the Progressive Era* (New York: Greenwood Press, 1988); Katherine Ott, *Fevered Lives: Tuberculosis in American Culture since 1870*

(Cambridge, MA: Harvard University Press, 1996); Barbara Bates, *Bargaining for Life: A Social History of Tuberculosis, 1876–1938* (Philadelphia: University of Pennsylvania Press, 1992). The few works that address children with bone tuberculosis include Richard Meckel, "Open Air Schools and the Tuberculous Child in Early Twentieth Century America," *Archives of Pediatric and Adolescent Medicine* 150 (1996): 91–96; Cynthia Connolly, *Saving Sickly Children: The Tuberculosis Preventorium in American Life, 1909–1970* (New Brunswick, NJ: Rutgers University Press, 2008); and Meghan Crnic and Cynthia Connolly "'They Can't Help Getting Well Here': Seaside Hospital for Children in the United States, 1872–1917," *Journal of the History of Childhood and Youth* 2 (2009): 220–33.

29. Michel Foucault, *Discipline and Punish: The Birth of the Prison,* trans. Alan Sheridan, 2nd ed. (New York: Vintage Books, 1995), 135ff. and 151–52. Foucault discusses how power relations in the modern nation-state became imprinted on the body. One way that the modern state disciplines the body is by controlling gesturing and physical posture. Indeed, in order to emphasize the importance of disciplined, upright bodies to the modern state, Foucault reprinted the frontispiece of orthopedist, Nicolas Andrey's *L'orthopédie*—an image of a crooked tree being pulled straight through stakes and twine (*Discipline and Punish,* fig. 10).

30. Little has been written specifically about the history of hernia care and repair. Ira Rutkow provides an overview of its early history in "A Selective History of Hernia Surgery in the Late Eighteenth Century: The Treatises of Percivall Pott, Jean Louis Petit, D. August Gottlieb Richter, Don Antonio de Gimbernat, and Pieter Camper," *Surgical Clinics of North America* 83, no. 5 (October 2003): 1021–44. The history of surgery is an interesting and thriving field of study. Some of the most notable books and articles include Lawrence, ed., *Medical Theory, Surgical Practice*; Pernick, *A Calculus of Suffering*; Thomas Schlich, *Surgery, Science and Industry: A Revolution in Fracture Care, 1950s–1990s* (Houndmills, Basingstoke: Palgrave, 2002); and Schlich, "The Emergence of Modern Surgery," in Deborah Brunton, ed., *Medicine Transformed: Health, Disease and Society in Europe, 1800–1930* (Manchester: Manchester University Press, 2004), 61–91.

31. Jesse T. Nicholson, "Founders of American Orthopedics," *Journal of Bone and Joint Surgery America* 481 (1966): 587.

32. For the history of US hospitals in the nineteenth century, see Charles E. Rosenberg, *The Care of Strangers: The Rise of America's Hospital System* (New York: Basic Books, 1987).

33. Philip D. Wilson, Jr., and David B. Levine, "Hospital for Special Surgery: A Brief Review of its Development and Current Position," *Clinical Orthopaedics and Related Research* 374 (May 2000): 90–106.

34. For nineteenth-century therapeutics in the United States, see John Harley Warner, *The Therapeutic Perspective: Medical Practice, Knowledge, and Identity in America, 1820–1885,* 2nd ed. (Princeton: Princeton University Press, 1997).

35. Royal Whitman, "Critical Estimation of the Personal Influence of Four Pioneers on the Development of Orthopaedic Surgery in New York," *Journal of Bone and Joint Surgery America* 16 (1934): 332.

36. For overviews of nineteenth-century orthopedic institutions in the United States, see Nicholson, "Founders of American Orthopedics"; Whitman, "A Critical Estimation"; Wilson and Levine, "Hospital for Special Surgery"; David Y. Cooper, "The Evolution of Orthopaedic Surgeons from Bone and Joint Surgery at the University of Pennsylvania," *Clinical Orthopaedics and Related Research* 374 (May 2000): 17–35; Henry J. Mankin, "Boston's Contributions to the Development of Orthopaedics in the United States," *Clinical Orthopeaedics and Related Research* 374 (May 2000): 47–54; and Brown, *The American Orthopaedic Association: A Centennial History.*

37. American orthopedists at the turn-of-the-twentieth century found Andry's definition of orthopedics too limited. In a letter to Elliot Brackett, Reginald H. Sayre wrote that Andry's book was little more than a "nursery guide," a book that was "in no sense a scientific work, even from the standpoint of that time." See Sayre, New York City to Elliot Brackett (Office of the Surgeon General, November 9, 1917, National Archives and Records Administration, College Park, Maryland [hereafter cited as NARA] II, RG 112, SGO 730, box 431). See also Royal Whitman, "A Definition of the Scope of Orthopedic Surgery," *Transactions of the American Orthopaedic Association* 8 (1896): 1–9.

38. Cooter, *Surgery and Society*, 11–12.

39. This comes from Henry E. Sigerist, *Medicine and Human Welfare* (New Haven: Yale University Press, 1941).

40. A more complete gender analysis of orthopedics and rehabilitation will be provided in the following chapter. For a history of motherhood, medicine, and the politics of blame, see Rima D. Apple, *Perfect Motherhood: Science and Childrearing in America* (New Brunswick, NJ: Rutgers University Press, 2006); Apple and Janet Golden, *Mothers and Motherhood: Readings in American History* (Columbus: Ohio State University Press, 1997); Molly Ladd-Taylor and Lauri Umansky, *"Bad" Mothers: The Politics of Blame in Twentieth-Century America* (New York: New York University Press, 1998).

41. Nicholson discusses this discovery in "Founders of American Orthopedics." See also Cooter, *Surgery and Society*, 13–18.

42. Nicholson, "Founders of American Orthopedics," 583.

43. Quoted in Nicholson, "Founders of American Orthopedics," 585.

44. Alfred R. Shands, "DeForest Willard, Philadelphia's Pioneer Orthopaedic Surgeon (1846–1910)," *Current Practice in Orthopedic Surgery* 4 (1969): 44.

45. From secondary sources, it appears that Willard was rejected because of his visual impairment. See Roland G. Curtin, "Memoir of DeForest Willard," *Transactions of the College of Physicians Philadelphia* 33 (October 4, 1911): 74–84.

46. See Curtin, "Memoir of DeForest Willard," and Shands, "DeForest Willard."
47. Charles Fayette Taylor, "The Spinal Assistant: Autobiographical Reminis- cences," *Transactions of the American Orthopaedic Association* 12 (1899): 19–20. Davis dictated these patient case histories to his daughter in 1887, shortly before his death.
48. See Lemuel F. Woodward, "Orthopedic Appliance Shop," *Transactions of the American Orthopaedic Association* 11 (1898): 76–82. For the history of sanatorium care, see footnote 27 above.
49. Taylor, "The Spinal Assistant."
50. Children are immunologically more likely to get Pott's disease. See Crnic and Connolly, "They Can't Help Getting Well Here," and Bradford and Lovett, *Treatise*, p. 10.
51. A. J. Steele quoted in Robert W. Lovett, "The Treatment of Pott's Disease," *Transactions of the American Orthopaedic Association* 8 (1896): 137–38.
52. Ibid., 138.
53. Lovett, "The Treatment of Pott's Disease," 136. At the time, Lovett was a senior member of the AOA—one year later he would be named president of the organization.
54. V. P. Gibney quoted in Lovett, "The Treatment of Pott's Disease," 139.
55. Reginald H. Sayre quoted in Lovett, "The Treatment of Pott's Disease," 149–50.
56. A. M. Phelps quoted in Lovett, "The Treatment of Pott's Disease," 147–48.
57. Ibid.
58. According to secondary sources on the history of respiratory tuberculosis, physicians did not worry about committing adults for long-term treatment, especially lower-class patients admitted to state institutions. For more on this, see Rothman, *Living in the Shadow of Death,* and especially Barron Lerner, *Contagion and Confinement: Controlling Tuberculosis Along the Skid Road* (Baltimore: Johns Hopkins University Press, 1998).
59. Phelps in Lovett, "The Treatment of Pott's Disease," 147–48.
60. Sayre in Lovett, "The Treatment of Pott's Disease," 149–50.
61. Because of the contagious nature of respiratory tuberculosis, different stan- dards were applied. See Lerner, *Contagion and Confinement.*
62. Katz, *In the Shadow of the Poorhouse,* 117–18.
63. Douglas C. McMurtrie, "The Care of Crippled Children in the United States," *American Journal of Orthopedic Surgery* 9 (May 1912): 527. For more on McMurtrie's role in the history of assisting children and adults with disabilities, see Brad Byrom, "A Pupil and a Patient: Hospital-Schools in Progressive America," in Paul Longmore and Lauri Umansky, eds., *The New Disability History: American Perspectives* (New York: New York University Press, 2001), 133–56.
64. Edith Reeves, *Care and Education of Crippled Children in the United States* (New York: Russell Sage Foundation, 1914), 5.

65. This number is based on McMurtie's listing of all such institutions in "The Care of Crippled Children in the United States."

66. See Katz, *In the Shadow of the Poorhouse,* and Linda Gordon, *The Great Arizona Orphan Abduction* (Cambridge, MA: Harvard University Press, 1999).

67. Reeves, *Care and Education of Crippled Children,* 22.

68. DeForest Willard, "Children's Orthopedic Ward, Agnew Memorial Pavilion," *Transactions of the American Orthopaedic Association* 11 (1898): 456–61.

69. DeForest Willard, "What Shall We Do With Our Cripples?" *Journal of the American Medical Association* 52 (April 3, 1909): 1134.

70. In order for a child to be admitted to the Widener, parents had to sign an "indenture" binding children over to the institution's trustees until the patients reached adulthood. For details of this arrangement, see Reeves, *Care and Education of Crippled Children,* 173–83. See also Willard, "What Shall we Do with our Cripples?" 1134.

71. Roger Cooter makes this same point; see *In the Name of the Child* and *Surgery and Society.*

72. Bradley Byrom also discusses the rise of hospital-schools in "A Vision of Self Support: Disability and the Rehabilitation Movement in Progressive America" (PhD diss., University of Iowa, 2004).

73. Gerald N. Grob, *The Mad among Us: A History of the Care of America's Mentally Ill* (New York: Free Press, 1994); Grob, *From Asylum to Community: Mental Health Policy in Modern America* (Princeton, NJ: Princeton University Press, 1991); and David J. Rothman, *The Discovery of the Asylum: Social Order and Disorder in the New Republic* (Boston: Little, Brown, 1971).

74. Harvey Kantor and David B. Tyack, "Historical Perspectives on Vocationalism in American Education," in *Work, Youth, and Schooling: Historical Perspectives on Vocationalism in American Education* (Stanford: Stanford University Press, 1982), 1. Other useful sources on the history of education and vocational training in the United States include: Michael Katz, *Reconstructing American Education* (Cambridge, MA: Harvard University Press, 1987), and Ira Katznelson, *Schooling for All: Class, Race, and the Decline of the Democratic Ideal* (New York: Basic Books, 1985).

75. For how the work ethic animated the poorhouse and many other corrective institutions in America, see Katz, *In the Shadow of the Poorhouse,* 12.

76. Reeves, *Care and Education,* 235.

77. Superintendent of New York's Charity Organization Society, Frank Persons, made these comments in response to Willard's paper. See Persons in Willard, "What Shall We Do with Our Cripples?" 1135.

78. Gwilym G. Davis, "The Relation of Cripples to the Public," *Transactions of the College of Physicians of Philadelphia* 39 (1917): 476.

79. For more on the interplay between eugenics and disability during the Progressive Era, see Martin Pernick, *The Black Stork: Eugenics and the Death of "Defective" Babies in American Medicine and Motion Pictures since 1915* (Oxford: Oxford University Press, 1996).

80. As described by Reeves, *Care and Education*, 151. Nurses and female educators played crucial roles in the day-to-day operations of hospital-schools such as the Widener. For more on the role of nursing in such institutions, see Connolly, *Saving Sickly Children*. See also Virginia Quiroga, "Female Lay Managers and Scientific Pediatrics at Nursery and Child's Hospital, 1854–1910," *Bulletin of the History of Medicine* 60 (1986): 192–208.

81. H. Winnett Orr, "The Industrial Education of the Crippled and Deformed," *American Journal of Orthopedic Surgery* 10 (1912): 195–200, 197. Edith Reeves who surveyed thirty-seven such institutions put it the following way: "little children, pale, worn and thin, and handicapped by disease or deformity, are found in [orthopedic industrial] school, proud of their ability to keep fully abreast of children in public schools . . . many of these children are eager to measure up to the accomplishments of normal children and become self-supporting, independent citizens"; Reeves, *Care and Education*, 3.

82. A similar division existed among sanatoriums devoted to respiratory tuberculosis. See Rothman, *Living in the Shadow of Death,* and Barron Lerner, *Contagion and Confinement.*

83. Orr, "The Industrial Education of the Crippled and Deformed," 199.

84. On the variations between states concerning worker's compensation, see Theda Skocpol, *Protecting Soldiers and Mothers: The Political Origins of Social Policy in the United States* (Cambridge, MA: Harvard University Press, 1992), 299ff. See also John Witt, *The Accidental Republic: Crippled Workingmen, Destitute Widows, and the Remaking of American Law* (Cambridge, MA: Harvard University Press, 2004), and Mark Aldrich, *Safety First: Technology, Labor, and Business in the Building of American Work Safety, 1870–1939* (Baltimore: Johns Hopkins University Press, 1997).

85. Gibney in Lovett, "The Treatment of Pott's Disease," 139–45.

86. Goldthwait interned at both the Boston Children's Hospital under the direction of Lovett and the Boston City Hospital under the tutelage of Bradford. J.G.K. "Joel Ernest Goldthwait, 1866–1961," *Journal of Bone and Joint Surgery* 43 (1961): 463–64. See also H. Winnett Orr, *Fifty Years of the American Orthopedic Association* (Lincoln, NB, 1937).

87. Joel E. Goldthwait, "With the Report of Eleven Cases Slipping or Recurrent Dislocation of the Patella," *American Journal of Orthopedic Surgery* 2 (1904): 293–308. The case was originally reported in "Dislocation of the Patella," *Transactions of the American Orthopaedic Association* 8 (1896): 237–38.

88. For weeding out unwanted workers, see Katz, *In the Shadow of the Poorhouse,* chap. 7. For "hopeless derelicts," see Joel E. Goldthwait, New York to Major Charles W. Mayo, Rochester, Minn, Oct. 8, 1917, NARA II, RG 112, SGO 730, box 431.

89. As quoted in Gritzer and Arluke, *The Making of Rehabilitation,* 43.

90. J.G.K, "Joel Ernest Goldthwait, 1866–1961."

91. See Beth Linker and Heather Perry, eds., *Globalizing Disability: World War I and the Making of Modern Rehabilitation,* forthcoming.

92. Joel Goldthwait, "The Place of Orthopedic Surgery in War," *American Journal of Orthopedic Surgery* 15 (1917): 679–86, and *The Division of Orthopaedic Surgery in the A.E.F.* (Norwood, MA: Plimpton Press, 1941).

93. Joel Goldthwait, "The Orthopedic Preparedness of the Nation," *American Journal of Orthopedic Surgery* 15 (1917): 219–20.

94. Timothy Dowling, ed., *Personal Perspectives: World War I* (Santa Barbara, CA: ABC-CLIO, 2006), 16.

95. Dowling, *Personal Perspectives,* especially the chapter titled " 'It's Only the Ones That Might Live Who Count': Allied Medical Personnel during World War I," 161–204.

96. John R. McDill, *Lessons from the Enemy; How Germany Cares for Her War Disabled* (Philadelphia and New York: Lea & Febiger, 1918).

97. Reznick, *Healing the Nation,* 119.

98. Cooter, *Surgery and Society,* 32ff.

99. See Koven "Remembering and Dismemberment," and Cooter, *Surgery and Society.*

100. M. Carter, "The Early Days of Orthopedic Nursing in the United Kingdom: Agnes Hunt and Baschurch," *Orthopaedic Nursing* 19 (2000): 15–18. See also Reznick, *Healing the Nation,* and Koven, "Remembering and Dismemberment."

101. Cooter, *Surgery and Society,* 114.

102. Prior to 1916, Shepherd Bush was funded exclusively by voluntary donations. For more on this see Reznick, *Healing the Nation,* 120ff.

103. As quoted in Reznick, *Healing the Nation,* 122.

104. Reznick, *Healing the Nation,* 123.

105. Cooter, *Surgery and Society,* 36ff.

106. As quoted in Goldthwait, *The Division of Orthopaedic Surgery,* 10.

107. Joel E. Goldthwait, London, UK, to the Office of the Surgeon General, Washington, D.C., July 16, 1917, NARA II, RG 112, SGO 730, box 431.

108. Samuel P. Hays, *Conservation and the Gospel of Efficiency: The Progressive Conservation Movement, 1890–1920* (Cambridge, MA: Harvard University Press, 1959): 3.

109. Goldthwait to the Office of the Surgeon General, July 16, 1917.

110. Goldthwait, *The Division of Orthopaedic Surgery,* 7.

111. "Orthopedic Advisory Council minutes," first meeting, August 12, 1917, as well as notes from the "Second Meeting of the Council," August 13, 1917, both found in NARA II, RG 112, SGO 730, box 431. For AOA presidencies, I am relying on the dates provided by Orr, *Fifty Years of the American Orthopedic Association.*

112. Gorgas, Office of the Surgeon General to Colonel Alfred E. Bradley, Department of Military Orthopedics, August 20, 1917, NARA II, RG 112, SGO 730, box 431.

113. This statement was found in a letter that Brackett wrote to the Council on National Defense. As quoted in Cooter, *Surgery and Society,* 129.

114. Goldthwait to Mayo, Oct. 8, 1917, NARA II, RG 112, SGO 730, box 431.

115. See Beth Linker, "No Scalpel Required: Orthopedic Surgeons and the 'Curative Workshop' in World War I America" (paper presented at the annual meeting for the American Association for the History of Medicine, Cleveland, Ohio, April 24, 2009). Since the inception of the AOA, orthopedic surgeons spent a great deal of time attempting to carve out an identity distinct from general surgery. One way that they did this was to define themselves as "conservatives" when it came to the knife. "Radical procedures characterize general surgery," Davis wrote in 1914, "whereas conservation is the watchword of the orthopedic surgeon." Conservative, in orthopedic circles, meant privileging mechanical manipulation over the scalpel. To be sure, it was vogue for surgeons of Davis's day to be conservatives. As medical historian Gert Brieger has pointed out, conservative surgery was an ideology employed by elite, university surgeons to separate themselves from their barber-surgeon past—to distance themselves from the image of the rash, careless, blood-thirsty hatchet men who recklessly chopped off limbs and cut out organs. The modern, scientific, conservative surgeon demonstrated restraint, precision, and educated judgment in his practice. But orthopedic surgeons took the notion of conservative surgery to its extreme, condemning the necessary condition that made operative procedures possible in the first place—namely, the use of the scalpel.

This did not mean that US orthopedic surgeons gave up performing surgical operations entirely. Most of them performed more operations in practice than they alluded to in their formal pronouncements on the matter. Nevertheless, US orthopedic surgeons wished to emphasize the necessity of mechanical treatment, treating operative work as secondary, inferior in both worth and status. Surgeons who treated orthopedic cases and privileged the knife were not considered to be true orthopedic surgeons. A case in point is that of New York City orthopedic surgeon Russell A. Hibbs, who in 1904 began treating Pott's disease with the knife, performing spinal fusions, surgically adhering the diseased vertebrae together. Because of Hibbs's preference for the knife over exercise and bracing, the leaders of the AOA denied him membership until well after the First World War. See Davis, "President's Address"; Gert Brieger, "From Conservative to Radical Surgery in Late Nineteenth-century America," in Lawrence, ed., *Medical Theory, Surgical Practice*, 216–31; and Brown, *The American Orthopaedic Association*.

116. Jennifer Keene, *Doughboys: The Great War, and the Remaking of America* (Baltimore: Johns Hopkins University Press, 2001).

117. Koven, "Remembering and Dismemberment," 1171.

CHAPTER THREE

1. Emphasis in the original. The creed appears in many issues of *Carry On: A Magazine on the Reconstruction of Disabled Soldiers and Sailors* and the

Publications of the Red Cross Institute for Crippled and Disabled Men. For the full creed, see *Carry On* 1, no. 9 (June 1919): 1.

2. For occupational therapy's prewar history, see Virginia A. M. Quiroga, *Occupational Therapy: The First Thirty Years, 1900–1930* (Bethesda, MD: American Occupational Therapy Association, 1995). A comprehensive historical account of dietetics still remains to be written. For a brief overview of dietitians during World War I, see Katherine E. Manchester and Helen B. Gearin, "Dieticians before World War II," in Robert S. Anderson, Harriet S. Lee, and Myra L. McDaniel, eds., *Army Medical Specialist Corps* (Washington, DC: Office of the Surgeon General, 1968), 15–39. See also Patricia A. M. Hodges, "Perspectives on History: Military Dietetics in Europe during World War I," *Journal of the American Dietetic Association* 93 (August 1993): 897–901.

3. Much of this analysis about physiotherapy and gender comes from Beth Linker, "Strength and Science: Gender, Physiotherapy, and Medicine in Early Twentieth Century America," *Journal of Women's History* 17, no. 3 (Fall 2005): 105–32. Copyright © *Journal of Women's History, Inc.* Reprinted with permission by The Johns Hopkins University Press. See also Ana Carden-Coyne, "Painful Bodies and Brutal Women: Remedial Massage, Gender Relations and Cultural Agency in Military Hospitals, 1914–18," *Journal of War and Culture Studies* 1, no. 2 (2008): 139–58.

4. Susan Jeffords, *The Remasculinization of America: Gender and the Vietnam War* (Bloomington: Indiana University Press, 1989).

5. See Karl Walter Hickel, "Entitling Citizens: World War I, Progressivism, and the Origins of the American Welfare State, 1917–1928" (PhD diss., Columbia University, 1999), 118.

6. Douglas McMurtrie, *The Organization, Work and Method of the Red Cross Institute for Crippled and Disabled Men* (New York: Red Cross Institute for Crippled and Disabled Men, 1918): 21. For more on sympathy and the medical profession, see selections from Ellen S. More and Maureen A. Milligan, eds., *The Empathic Practitioner: Empathy, Gender, and Medicine* (New Brunswick, NJ: Rutgers University Press, 1994).

7. For more on the Progressive Era movement to make charity scientific, see Emily K. Abel, "Valuing Care: Turn-of-the-Century Conflicts between Charity Workers and Women Clients," *Journal of Women's History* 10, no. 3 (Autumn 1998): 32–52.

8. See "Second Meeting of the Orthopedic Advisory Council," NARA II, RG 112, SGO 730, box 430.

9. See Goldthwait, New York, N.Y. to Major Charles W. Mayo, Rochester, Minn., Oct. 8, 1917, NARA, II, RG 112, SGO 730, box 430.

10. See Joel E. Goldthwait, *The Division of Orthopaedic Surgery in the A.E.F.* (Norwood, MA: Plimpton Press, 1941), esp. 16–19 and 109–13. In a letter to Frank Billings in February 1919, Goldthwait wrote, "as I wrote you in a previous letter, all of the [occupational therapists] are really, when analyzed,

Physio-therapeutic Aides, and for that reason are being so considered"; see Goldthwait, pp. 112–13. For more on Jones's system of rehabilitation, see Roger Cooter, *Surgery and Society in Peace and War: Orthopaedics and the Organization of Modern Medicine, 1880–1948* (London: Macmillan Press, 1993), and Jeffrey S. Reznick, *Healing the Nation: Soldiers and the Culture of Caregiving in Britain during the Great War* (Manchester: Manchester University Press, 2004).

11. Goldthwait also praises Britain's occupational therapy program, where women instructed maimed soldiers in crocheting, knitting, embroidery, as well as industrial activities, such as metal work, leatherwork, stenography, typing, and machine work. Basically, occupational therapists developed programs of work that would require soldiers to use their damaged limbs. Surgeons like Goldthwait believed that such therapy would have both curative and vocational benefits. Goldthwait was so impressed with the program that upon return to the United States, he helped his wife start a school for occupational therapists based out of their home in Boston. Many of Goldthwait's graduates later assisted him in the US base hospitals in France. For Goldthwait's personal account, see *Division of Orthopaedic Surgery*, 16–20, 109–13. It is important to note that at the time Goldthwait was in Britain, he was ostensibly unaware of a small contingent of women in the United States already involved in occupational therapy. For more on the pre–World War I occupational therapy movement, see Quiroga, *Occupational Therapy*.

12. Most therapists working for the Royal Medical Corps were masseuses, either from the exclusively female professional organization of the Society of Trained Masseuses (est. 1894) or the male-dominated Institute of Massage and Remedial Gymnastics (est. 1916). For British physiotherapy during World War I, see Jean Barclay, *In Good Hands: The History of the Chartered Society of Physiotherapy, 1894–1994* (Oxford: Butterworth-Heinemann, 1994), and Gerald Larkin, *Occupational Monopoly and Modern Medicine* (London: Tavistock Publications, 1983), 92–124. Ruby Heap, who studies the history of physiotherapy in Canada, also gives some details about the British system in, "'Salvaging War's Waste': The University of Toronto and the 'Physical Reconstruction' of Disabled Soldiers during the First World War," in Edgar-André Montigny and Lori Chambers, eds., *Ontario since Confederation: A Reader* (Toronto: University of Toronto Press, 2000), 224. See also, Cooter, *Surgery and Society*, 105.

13. Heap, "Salvaging War's Waste," 224–25.

14. The Sanitary Commission was a quasi-governmental organization during the Civil War that disappeared afterwards; the impetus and interested parties were absorbed into the Red Cross movement. The Sanitary Corps was a male-dominated department that included pharmacists, sanitary engineers, and microbiologists. For more details, see Richard V. N. Ginn, *The History of the U.S. Army Medical Service Corps* (Washington, DC: Office of the Surgeon

General and Center of Military History, United States Army, 1997). For an account of the Sanitary Commission, see Judith Ann Giesberg, *Civil War Sisterhood: The U.S. Sanitary Commission and Women's Politics in Transition* (Boston: Northeastern University Press, 2000).

15. See "Second Meeting of the Orthopedic Advisory Council," NARA II, RG 112, SGO 730, box 430.

16. See Heap, "Salvaging War's Waste," 225. See also Barclay, *In Good Hands*.

17. A few exceptions existed, such as Henry O. Kendall. For the several men who were trained as physiotherapists during the First World War, see Linker, "Strength and Science."

18. *MDWW*, vol. 13, pt. 1 (Washington, DC: GPO, 1927), 57.

19. By the First World War, nursing had become fully institutionalized within the Army Medical Department. In 1901 the military created the Army Nurses Corps, granting nurses full commission. This meant that nurses had military status, giving them similar insurance and housing benefits granted to all entering soldiers. Physiotherapists would not receive military status in the United States Army until 1947.

20. See A. B. Hirsch, M.D., to Dr. Richard Kovacs, New York City, July 1, 1925, American Physical Therapy Association [APTA] archives, BOD 1, file 14. For history of nursing during this era, see Patricia D'Antonio, *American Nursing: A History of Knowledge, Authority, and the Meaning of Work* (Baltimore: Johns Hopkins University Press, 2010), and Susan Reverby, *Ordered to Care: The Dilemma of American Nursing, 1850–1945* (Cambridge: Cambridge University Press, 1987). For the history of military nursing, see Christine Hallett, "The Personal Writings of First World War Nurses: A Study of the Interplay of Authorial Intention and Scholarly Interpretation," *Nursing Inquiry* 14 (2007): 320–29.

21. Louisa Lippitt, R.N., to Dr. Frank Granger, Madison, Wisconsin, February 19, 1920, APTA archives, BOD 1, file 1. For Granger's response that nurses were not good skilled operators, see Frank Granger to Louisa Lippitt, R.N., Boston, February 26, 1920, APTA archives, BOD 1, file 1.

22. Douglas McMurtrie, *The Disabled Soldier* (New York: Macmillan, 1919), 19.

23. Douglas McMurtrie, "War-Cripple Axioms and Fundamentals," *Columbia University War Pamphlet* 17 (November 17, 1917): 65.

24. APTA archives, box 136, file 24, 1. Informal talk given by Alice Lou Plastridge, November 2, 1974, at a dinner in honor of five life members of the American Physical Therapy Association.

25. Plastridge, 1. See also Wendy Murphy, *Healing the Generations: A History of Physical Therapy and the American Physical Therapy Association* (Alexandria: American Physical Therapy Association, 1995), 34–35, and anonymous, "Infantile Paralysis: Pioneers in Treatment," *Physical Therapy* 56 (1976): 42. See also Naomi Rogers, *Dirt and Disease: Polio before FDR* (New Brunswick, NJ: Rutgers University Press, 1992), and David M Oshinsky, *Polio: An American Story* (Oxford: Oxford University Press, 2005).

26. The historical record on Sanderson is extremely scant, partly because after she got married in 1922, she ostensibly gave up her career in physical education and physiotherapy entirely. The little information that I could glean about her career came from "News Notes," *American Physical Education Review* 20 (January 1915): 42–43; Emma Vogel, "Physical Therapists before WWII," in Robert S. Anderson, Harriet S. Lee, and Myra L. McDaniel, eds., *Army Medical Specialist Corps* (Washington, DC: Office of the Surgeon General, 1968), 44–57, and "Over the Top with Miss Sanderson and the First Overseas Unit," *P.T. Review* 2 (1922): 3–4.
27. Vogel, "Over the Top," 3–4.
28. See Vogel, "Physical Therapists," 44.
29. Murphy, *Healing the Generations,* 50.
30. Marguerite Irvine, "Recollections and Reminiscences from Former Reconstruction Aides," *Physical Therapy* 56 (January 1976): 29. See also Lettie Gavin, *American Women in World War I: They Also Served* (Niwot: University of Colorado Press, 1997): 101–28.
31. Doris Ball Crawford, "Recollections and Reminiscences from Former Reconstruction Aides," *Physical Therapy* 56 (January 1976): 26.
32. These are just three verses of a seven-verse song that the Reed College graduates composed. See APTA archives, box 6, file 1.
33. Historian Lucy Bland has characterized the struggle of women professionals and their sexuality the following way: "Behind the veneer of the dominant nineteenth-century ideal woman—the domestic 'angel in the house'—lurked the earlier representation of sexualized femininity: The Magdalene behind the Madonna"; Bland, *Banishing the Beast: Feminism, Sex and Morality* (New York: New Press, 1995): 58. See also D. A. Nicholls and J. Cheek, "Physiotherapy and the Shadow of Prostitution: The Society of Trained Masseuses and the Massage Scandals of 1894," *Social Science and Medicine* 62, no. 9 (2006): 2339.
34. Hartwig Nissen, "The Swedish Movement and Massage Treatment," *Journal of the American Medical Association* 10, no. 14 (April 7, 1888): 423–24.
35. "The Scandals of Massage: Report on the Special Commissioners of the British Medical Journal," *British Medical Journal* 2, no. 1769 (Nov. 24, 1894): 1199–1200.
36. Nicholls and Cheek, "Physiotherapy and the Shadow of Prostitution," 2342.
37. For "young men about town," see "Immoral 'Massage' Establishments," *British Medical Journal* 2, no. 1750 (July 14, 1894): 88.
38. Larkin, *Occupational Monopoly and Modern Medicine,* also makes this point.
39. The most thorough account of the CTCA is Nancy K. Bristow, *Making Men Moral: Social Engineering During the Great War* (New York: New York University Press, 1996). See also, Allan M Brandt, *No Magic Bullet: A Social History of Venereal Disease in the United States Since 1880* (New York: Oxford University Press, 1985), and Alexandra M. Lord, " 'Naturally Clean and

Wholesome': Women, Sex Education, and the United States Public Health Service, 1918–1928," *Social History of Medicine* 17, no. 3 (December 1, 2004): 423–41.

40. Brandt, *No Magic Bullet*, 66–67.

41. Before taking his post at Johns Hopkins, Nissen worked at Wellesley's physical education program, teaching the all-women study body in techniques of Swedish exercises and massage. See Martha Verbrugge, *Able-Bodied Womanhood: Personal Health and Social Change in Nineteenth-Century Boston* (Oxford: Oxford University Press, 1988): 151.

42. See Martha Verbrugge, "Knowledge and Power: Health and Physical Education for Women in America," in Rima D. Apple, ed., *Women, Health, and Medicine in America: A Historical Handbook* (New Brunswick: Rutgers University Press, 1992), 361–82.

43. Paul Atkinson, "The Feminist Physique: Physical Education and the Medicalization of Women's Education," in J. A. Mangan and Roberta J. Park, eds., *From 'Fair Sex' to Feminism: Sport and the Socialization of Women in the Industrial and Post-Industrial Eras* (London: Frank Cass & Co., 1987), 38–57. In the same volume, see also Carroll Smith-Rosenberg and Charles Rosenberg, "The Female Animal: Medical and Biological Views of Women and their Role in Nineteenth-century America," 13–37; and Patricia Vertinsky, "Body Shapes: The Role of the Medical Establishment in Informing Female Exercise and Physical Education in Nineteenth-century North America," 256–81.

44. Murphy, *Healing the Generations*, 12.

45. Nicholls and Cheek, "Physiotherapy and the Shadow of Prostitution," 2343.

46. Gavin, *American Women in World War I*, 107.

47. Lena Hitchcock, "The Great Adventure" (unpublished memoir), iii and vii, National World War I Museum at Liberty Memorial, Kansas City, Missouri.

48. Hitchcock, "The Great Adventure," 8.

49. Quiroga, *Occupational Therapy*, 35–42.

50. For "Jane Addams of occupational therapy," see Quiroga, *Occupational Therapy*, 35. See also T. J. Jackson Lears, *No Place of Grace: Antimodernism and the Transformation of American Culture, 1880–1920*, 2nd ed. (Chicago: University of Chicago Press, 1994), 60–96.

51. "Ever Play It?," *P.T. Review* 1 (December 1921): 4.

52. Hitchcock, "The Great Adventure," 1.

53. See Walsh, "The Administration and Planning of the Hospital Physical Therapy Department," delivered at the Annual Congress on Medical Education, Medical Licensure and Hospitals, Chicago, Illinois, February 18, 1930, APTA archives, box 6, file 6.

54. As quoted in Goldthwait, *Division of Orthopaedic Surgery*, 113.

55. Ruby Decker, "Recollections and Reminiscences from Former Reconstruction Aides," *Physical Therapy* 56 (January 1976): 25.

56. For case record, see *MDWW*, vol. 13, pt. 1, 271, 277.

57. Ibid., 124.
58. For specs, see ibid., 16. For description of Fort McPherson, see ibid., 124ff.
59. Ibid., 126.
60. Lydia Dock, "Is the Profession Becoming Overcrowded?" *American Journal of Nursing* 3 (April 1903): 513–15. See also L. L. Dock, Anna E. Rutherford, and Helen Kelly, "Letters to the Editor," *American Journal of Nursing* 2, no. 1 (1901): 62–66; and Helen Conkling Bartlett, "The Teaching of Massage to Pupils in Hospital Training-Schools," *American Journal of Nursing* 1, no. 10 (1901): 718–21.
61. Marguerite Sanderson, "Some Aspects of the Massage Problem," *American Journal of Orthopedic Surgery* 16 (1918): 437.
62. Mary McMillan, *Massage and Therapeutic Exercise* (Philadelphia: W. B. Saunders Company, 1921).
63. In regular camp, military authorities believed that organized athletics could combat problems of desertion, alcohol, and the lure of prostitution. They also thought that military sport would enhance the "physical manhood" of US draftees, especially since in some areas of the country as many as a third of all draftees had been rejected from service because of physical "defects" and weakness. The military leaders in charge of pre-combat training camps designed sporting events that would teach soldiers "necessary survival skills for life on the front." Baseball, it was argued, was useful in preparing men to fight a war, since the skills needed for pitching were similar to those for grenade tossing. Likewise, gymnastic exercises on balance beams incorporated abilities necessary for "daily trench maneuvers." boxing and wrestling matches, however, proved to be the most popular at basic training camps, even though prior to 1917 boxing was illegal in most states. US sergeants who taught recruits bayonet exercises liked boxing in particular, argued that the "essential movement of feet, hands, and body in bayonet fighting are the same as those in boxing." Steven W. Pope, "An Army of Athletes: Playing Field, Battlefields, and the American Military Sporting Experience, 1890–1920," *Journal of Military History* 59, no. 3 (July 1995): 435–56.
64. Verbrugge, "Recreating the Body: Women's Physical Education and the Science of Sex Differences in America, 1900–1940," *Bulletin of the History of Medicine* 71 (1997): 273–304, 278.
65. As quoted in Verbrugge, "Recreating the Body," 286.
66. This variation can best be seen when comparing visual images of amputees playing baseball in magazine such as *Carry On* with those from the film *Heroes All.*
67. Verbrugge, "Recreating the Body."
68. There are many places in the primary literature on this topic that speaks of sport and physical education in terms of virility-building. One place to see how the argument is sustained is "Wounded Must Build Vitality: Lieut. Williams Tells Men how to Come Back Physically," *The Come-Back,* January 24, 1918, 2.

69. Rehabilitators claimed that disabled soldiers needed sporting events in order to fully recuperate. Baldwin worried about motivation, about how to get disabled soldiers to engage in all facets of rehabilitation, from physical therapy to taking up the simulated work provided by the curative workshops. He concluded that soldiers needed to adopt a competitive spirit and that the key ingredient to getting disabled soldiers to comply with rehabilitation was competition. He developed range of motion devices (similar in kind to devices used daily by physical therapists today) that measured joint mobility, the results of which were made visible both to the treating therapist and the patient. *MDWW,* vol. 13, pt. 1, 113.

70. "Those Lady Athletes Keep Things Hummin': Reconstruction Aides, All Chesty, Challenge Nurses for Basketball," *The Come-Back,* March 19, 1919, 5.

71. For a personal account of the labor required to perform physiotherapy, see Ida May Hazenhyer, "A History of the American Physiotherapy Association," *Physiotherapy Review* 26 (February 1946): 3–14, 66–74, 122–29, 174–84.

72. See caricatures and pictorial depictions in the *P.T. Review* 1 (March 1921), reprinted in Hazenhyer, "A History of the American Physiotherapy Association," 9.

73. See Patricia A. M. Hodges, "Perspectives on History: Military Dietetics in Europe during World War I," *Journal of the American Dietetic Association* 93 (August 1993): 897–901, 900.

74. Over the course of the 1920s and 1930s, APA therapists shed the feminine characteristics of their journal (such as birth and wedding announcements), transforming it into a more streamlined, objective publication. The issue of purging themselves of a feminine, maternalistic image came to a head when they drafted their first professional code of ethics and promised that their organization's creed would avoid any hint of "sentimentalism." For more on this, see Beth Linker, "The Business of Ethics: Women, Medicine, and the American Physiotherapy Association's 1935 Code of Ethics," *Journal of the History of Medicine and Allied Sciences* 60 (July 2005): 321–54. By permission of Oxford University Press. For a reprint of the APA's 1935 code of ethics, see Ruth Purtilo, "The American Physical Therapy Association's Code of Ethics: Its Historical Foundations," *Physical Therapy* 57 (September 1977): 1001–6.

75. "Physio-Therapy's Part in Reconstruction," *Carry On* 1 (June 1919): 7–9.

76. Alice Duer Miller, "How Can a Woman Best Help," *Carry On* 1 (June 1918): 17–18.

77. See the cover art of *Carry On* 1 (July 1919).

CHAPTER FOUR

1. Sanders Marble, *Rehabilitating the Wounded: Historical Perspective on Army Policy* (Washington, DC: Office of Medical History, 2008), 12, 21.

2. Memo of November 7, *MDWW*, vol. 13, pt. 1 (Washington, DC: GPO, 1927), 9–10.
3. Marble, *Rehabilitating the Wounded*, 3.
4. For "burden on the State," see World War I orthopedic surgeon, Fred H. Albee, "The Function of the Military Orthopedic Hospital," *New York Medical Journal* 106 (1917): 2. My argument here diverges from Rosemary Stevens's argument that one of the major characteristics of American hospitals at this time was their "focus on acute care"; see *In Sickness and in Wealth: American Hospitals in the Twentieth Century*, 2nd ed. (Baltimore: Johns Hopkins University Press, 1999), 11. In many ways the early twentieth-century development of military hospitals was distinct from that of civilian hospitals. Most significantly, the army, through its system of military hospitals, felt compelled to meet the growing demands to overhaul the veteran welfare policy.
5. "14 Hospitals Chosen for War's Disabled," *New York Times*, April 1, 1918, 7.
6. Most scholars who study the rise of the "modern hospital" in the United States—an institution that, rather reflexively, historians often date as beginning with the First World War—list surgery, childbirth, and various diagnostic technologies as the most influential factors leading to its development. Although World War I serves as a significant event distinguishing the modern from the premodern, little work has been done on what role the war itself played in the development of the modern hospital. For more on the rise of the modern hospital in the United States, see Charles E. Rosenberg, *The Care of Strangers: The Rise of America's Hospital System* (New York: Basic Books, 1987); Rosemary Stevens, *In Sickness and In Wealth*; Joel D. Howell, *Technology in the Hospital: Transforming Patient Care in the Early Twentieth Century* (Baltimore: Johns Hopkins University Press, 1995); Diana E. Long and Janet Golden, eds., *The American General Hospital: Communities and Social Contexts* (Ithaca, NY: Cornell University Press, 1989); Morris J. Vogel, *The Invention of the Modern Hospital: Boston 1870–1930* (Chicago: University of Chicago Press, 1980); and David Rosner, *A Once Charitable Enterprise: Hospitals and Health Care in Brooklyn and New York, 1885–1915* (Cambridge and New York: Cambridge University Press, 1982). For how changes in childbirth practice shaped the modern hospital, see Judith Walzer Leavitt, *Brought to Bed: Childbearing in America, 1750 to 1950* (New York: Oxford University Press, 1986). For the effects of surgery, see Daniel M. Fox and Christopher Lawrence, *Photographing Medicine: Images and Power in Britain and America since 1840* (New York: Greenwood Press, 1988), as well as Christopher Lawrence, ed., *Medical Theory, Surgical Practice: Studies in the History of Surgery* (London: Routledge, 1992).
7. Edward Stevens, "The Physiotherapy Department for the American Hospital," *Architectural Record* 60 (July 1926): 18–24.
8. Leo Mayer, "The Organization and Aims of the Orthopedic Reconstruction Hospital," *American Journal of the Care for Cripples* 5 (September 1917): 84.

9. Edward Rich, Letterman General Hospital, California to Elliot B. Brackett, Surgeon General's Office, Washington, DC, July 8, 1918, NARA II, RG 112, "SGO, 1917–1927, Letterman," box 101, file 721.1.

10. See L. C. Mudd, "The Letterman General Hospital," *New York Medical Journal* 109 (1919): 242–44.

11. W. F. Southard, "Opening of the New Military Hospital at the Presidio," *Pacific Medical Journal* 42 (July 9, 1899): 454–64.

12. Annmarie Adams, *Medicine by Design: The Architect and the Modern Hospital, 1893–1943* (Minneapolis: University of Minnesota Press, 2008): 9.

13. See John Elliott Brown and Edward F. Stevens, "General Hospital for One Hundred Patients," in Charlotte A. Aikens, ed., *Hospital Management* (Philadelphia: Saunders, 1911), 108–47.

14. Southard, "Opening of the New Military Hospital."

15. John D. Thompson and Grace Goldin, *The Hospital: A Social and Architectural History* (New Haven: Yale University Press, 1975), 170ff.

16. William C. Borden, "The Walter Reed General Hospital of the United States Army," *Military Surgeon* 20 (1907): 23.

17. William Bennett Bean, *Walter Reed: A Biography* (Charlottesville: University Press of Virginia, 1982), and John R. Pierce, *Yellow Jack: How Yellow Fever Ravaged America and Walter Reed Discovered Its Deadly Secrets* (Hoboken, NJ: J. Wiley, 2005). For another view on Walter Reed, see Mariola Espinosa, *Epidemic Invasions: Yellow Fever and the Limits of Cuban Independence* (Chicago: University of Chicago Press, 2009).

18. Borden, "The Walter Reed General Hospital," 33.

19. W. O. Owen, "The Army Medical Museum," *New York Medical Journal* 107 (June 1918): 1034–36, and Michael Rhode, "Photography and the Army Medical Museum, 1862–1945," *Architext* 4, no. 2 (March 1995): 7–10. See also P. M. Ashburn, *A History of the Medical Department of the United States Army* (Boston and New York: Houghton Mifflin Company, 1929).

20. Sherman Fleek, " 'Borden's Dream' Leads to Walter Reed Innovation Century Ago," US Army, http://www.army.mil/-news/2009/05/01/20472-bordens-dream-leads-to-walter-reed-innovation-century-ago/.

21. The history of Walter Reed Hospital, especially in relation to the Army Medical School, the Army Medical Museum (now known as the National Museum of Health and Medicine), and the Surgeon General's Library (now the National Library of Medicine) is complex. The best single source is Mary W. Standlee, *Borden's Dream: The Walter Reed Army Medical Center in Washington, D.C.* (Washington, DC: Borden Institute, 2009).

22. Borden, "The Walter Reed General Hospital," 26.

23. Ibid., 30.

24. Ibid., 25.

25. Ibid., 27.

26. Southard, "Opening of the New Military Hospital." 458.

27. Borden, "The Walter Reed General Hospital," 28.

28. *MDWW* 5: 311.
29. Bradley Allen Byrom, "A Vision of Self Support: Disability and the Rehabilitation Movement in Progressive America" (PhD diss., University of Iowa, 2004); see also his "A Pupil and a Patient: Hospital-Schools in Progressive America," in Paul K. Longmore and Lauri Umansky eds., *The New Disability History: American Perspectives* (New York: New York University Press, 2001), 133–56.
30. Lt. C's account can be found in Fred H. Albee, *A Surgeon's Fight to Rebuild Men: An Autobiography* (New York: Dutton, 1943), 170–72. For a history of the Carrel-Dakin treatment, see Perrin Selcer, "Standardizing Wounds: Alexis Carrel and the Scientific Management of Life in the First World War," *British Journal for the History of Science* 41 (2008): 73–107.
31. Albee, *A Surgeon's Fight to Rebuild Men.*
32. *MDWW* 5: 306.
33. These numbers have been gleaned from the monthly memos that the Letterman orthopedic service sent to the Army Surgeon General's Office from 1918 to 1919. See Major Robert L. Hull, Letterman to OSG, September, 1918; Major Leo Eloesser, Letterman to OSG, Dec. 1918 and Major Eloesser, Letterman, to OSG, Feb. 1, 1919, all found in NARA II, RG 112, "SGO 1917–1927, Letterman," box 101, file 721.1.
34. *MDWW* 5: 283.
35. Mayer, "The Organization and Aims," 79–80.
36. *The History of Letterman General Hospital* (San Francisco, CA: Listening Post, 1919), 24. This was the first war in US history in which a diagnosis of flat feet could preclude a man from service. The US Medical Department enlisted orthopedic surgeons to rehabilitate draftees with flat feet to make them physically fit for military duty. For more on this history, see Beth Linker, "Feet for Fighting: Locating Disability and Social Medicine in World War I America," *Social History of Medicine* 20 (2007): 91–109.
37. C. R. Darnall to Emmett J. Scott, Washington, D.C., Memorandum, 19 December 1918, NARA I, RG 15, box 254. Darnall wrote to Scott that "disabled soldiers will be sent to the General and Base Hospitals without regard to color."
38. As quoted in Mark Ellis, *Race, War, and Surveillance: African Americans and the United States Government during World War I* (Bloomington: Indiana University Press, 2001), 56.
39. Emmett J. Scott, *The American Negro in the World War*, 2nd ed. (New York: Arno Press, 1969), 447. Scott's original history of black involvement in World War I was first published in 1919.
40. Ibid. Some of Scott's contemporaries as well as historians who study the period argue that little change happened in the War Department and the Surgeon General's Office because of Scott's accommodationist background on issues of race. Scott was a Tuskegee loyalist and served as Booker T. Washington's secretary prior to his position in the War Department. For

more on this and Scott's relationship to the NAACP and to W. E. B. DuBois, see Mark Ellis, *Race, War, and Surveillance,* and Maceo Crenshaw Daily, "Neither 'Uncle Tom' nor 'Accommodationist': Booker T. Washington, Emmett Jay Scott, and Constructionalism," *Atlanta History* 38, no. 4 (1995): 20–33.

41. Scott, *The American Negro,* 429.

42. C. R. Darnall, Washington, D.C. to Emmett J. Scott, Special Assistant to Secretary of War, War and Navy Building, Washington, D.C., December 19, 1918, NARA I, RG 15, box 254.

43. Ibid., and Scott, *The American Negro.* The best treatment of black-white segregation in hospitals during the twentieth century is Vanessa Northington Gamble, *Making a Place for Ourselves: The Black Hospital Movement, 1920–1945* (New York: Oxford University Press, 1995). For other histories of black medical care, see Ellis, *Race, War, and Surveillance,* and Edward H. Beardsley, *A History of Neglect: Health Care for Blacks and Mill Workers in the Twentieth-Century South* (Knoxville: University of Tennessee Press, 1987). Histories of tuberculosis and sanatoriums of the twentieth century have also addressed the inequalities in medical care based on race. See, for instance, David McBride, *From TB to AIDS: Epidemics among Urban Blacks since 1900* (Albany: State University of New York Press, 1991) and Katherine Ott, "Raceing Illness at the Turn of the Century," in *Fevered Lives: Tuberculosis in American Culture since 1870* (Cambridge, MA: Harvard University Press, 1996), 100–110.

44. For a general overview of what a typical orthopedic appliance shop would include, see Lemuel F. Woodward, "Orthopedic Appliance Shop," *Journal of Bone and Joint Surgery* 11 (1898): 76–82. See also Albee, "The Function of the Military Orthopedic Hospital," 2–4.

45. See Robert B. Osgood, Letterman General Hospital to the Office of the Surgeon General, Washington, D.C., August 29, 1918; Robert L. Hull, Letterman General Hospital to the Office of the Surgeon General, Washington, D.C., September, 1918; and Leo Eloesser, Letterman General Hospital to the Office of the Surgeon General, Washington, D.C., December 1918, NARA II, RG 112, "SGO 1917–1927, Letterman," box 101, file 721.1.

46. *MDWW,* vol. 13, pt. 1, 16.

47. *MDWW* 5: 284. Rehabilitation was not the only reason for the power plant. Electricity was needed to heat and light the ever-expanding number of hospital wards as well.

48. *MDWW* 5: 57.

49. *MDWW,* vol. 13, pt. 1, 271.

50. *MDWW* 5: 311.

51. *The History of Letterman General Hospital,* 24.

52. Major Robert L. Hull, Letterman General Hospital to the Army Surgeon General's Office, September 1918, NARA II, RG 112, "SGO 1917–1927, Letterman," box 101, file 721.1.

53. The psychology department at Walter Reed designed joint range-of-motion measuring devices (instruments primarily used by physical therapists today) for the dual purpose of recording patient progress and motivating patients to beat their own scores as well as the scores of their comrades. For more on the Walter Reed psychology department see Bird T. Baldwin, *Occupational Therapy Applied to Restoration of Movement* (Washington, DC: Walter Reed General Hospital, 1919).

54. "Walter Reed Hospital Annual Report, 1918," found in NARA II, R.G. 112, "SGO 1917–1927, Walter Reed," box 148, file 391.

55. Leo Mayer, "The Military Orthopedic Reconstruction Hospital," *Journal of the American Medical Association* 69, no. 18 (November 3, 1917): 1523.

56. "Walter Reed Hospital Annual Report, 1918," found in NARA II, R.G. 112, "SGO 1917–1927, Walter Reed," box 148, file 391.

57. Work therapy proved to be more common in Great Britain than in the United States. See, for instance, Linda Bryder, *Below the Magic Mountain: A Social History of Tuberculosis in Twentieth-Century Britain* (Oxford: Clarendon Press, 1988). Because of the influence of S. Weir Mitchell, US tuberculosis physicians favored the rest cure. For more on this, see Beth Linker, "Rest Cure," in *The Encyclopedia of American Disability History,* ed. Susan Burch (New York: Facts on File, 2009), 3: 780–81.

58. For more on Paterson, see R. Y. Keers, *Pulmonary Tuberculosis: A Journey down the Centuries* (London: Bailliere Tindall, 1978).

59. Gerald N Grob, *The Mad among Us: A History of the Care of America's Mentally Ill* (New York: Free Press, 1994), 27. See also Patricia D'Antonio, *Founding Friends: Families, Staff, and Patients at the Friends Asylum in Early Nineteenth-Century Philadelphia* (Bethlehem, PA: Lehigh University Press, 2006).

60. For a thorough account of Kirkbride's professional life, see Nancy Tomes, *A Generous Confidence: Thomas Story Kirkbride and the Art of Asylum-Keeping, 1840–1883* (Cambridge: Cambridge University Press, 1984).

61. Beth Haller and Robin Larsen, "Persuading Sanity: Magic Lantern Images and the Nineteenth-Century Moral Treatment in America," *Journal of American Culture* 28, no. 3 (2005): 259–72.

62. Philippe Pinel, *A Treatise on Insanity* (Sheffield: Todd, 1806), 113, quoted in Grob, *Mad among Us,* 27.

63. Sir Robert Jones of Great Britain also preferred work therapy, which he called "indirect therapy." For more on this see Jeff Reznick, *Healing the Nation: Soldiers and the Culture of Caregiving in Britain During the Great War* (Manchester: Manchester University Press, 2004), 127ff.

64. *The History of Letterman General Hospital,* 16. Joel Goldthwait, director of orthopedics for the AEF, was of like mind. As he wrote: "That 'Satan finds mischief for idle hands' was never more correctly demonstrated than with these large numbers of wounded men placed in the hospitals with nothing to do while the healing was going on but brood over the things that had

happened to them"; Goldthwait, *The Division of Orthopaedic Surgery in the A.E.F.* (Norwood, MA: Plimpton Press, 1941), 111.

65. See Albee, *A Surgeon's Fight,* 199.

66. Annmarie Adams, "Borrowed Buildings: Canada's Temporary Hospitals during World War I," *Canadian Bulletin of the History of Medicine* 16 (1999): 44.

67. Patrick Kelly points out that, as opposed to National Homes located on the East Coast, in Cleveland, and in Chicago, "female benevolent workers enjoyed almost complete autonomy in governing and managing local homes created for Union veterans"; see *Creating a National Home: Building the Veterans' Welfare State, 1860–1900* (Cambridge, MA: Harvard University, 1997), 35.

68. For Frederick Knapp, see Kelly, *Creating a National Home,* 49–50, 155.

69. Reznick makes a similar point about how Sir Robert Jones wanted his soldier-patients to become "self-healing machines"; see *Healing the Nation,* 127.

70. *The History of Letterman General Hospital,* 6.

71. As quoted in Kelly, *Creating a National Home,* 73.

72. Kelly, *Creating a National Home,* 68, 25, and 13.

CHAPTER FIVE

1. For German casualty statistics, see Robert Gerald Whalen, *Bitter Wounds: German Victims of the Great War, 1914–1939* (Ithaca, NY: Cornell University Press, 1984), 40, and Heather Perry, "Re-Arming the Disabled Veteran: Artificially Rebuilding State and Society in World War I Germany," in Katherine Ott, David Serlin, and Stephen Mihm, eds., *Artificial Parts, Practical Lives: Modern Histories of Prosthetics* (New York: New York University Press, 2002), 78.

2. Joanna Bourke, *Dismembering the Male: Men's Bodies, Britain and the Great War* (London: Reaktion Books, 1996). For numbers of "permanently disabled," see Deborah Cohen, *The War Come Home: Disabled Veterans in Britain and Germany, 1914–1939* (Berkeley: University of California Press, 2001), 1–4 and appendix. For numbers of amputees, see Mary Guyatt, "Better Legs: Artificial Limbs for British Veterans of the First World War," *Journal of Design History* 14 (2001): 311.

3. Robert B. Osgood, "A Survey of the Orthopaedic Services in the US Army Hospitals, General Base, and Debarkation," *American Journal of Orthopedic Surgery* 17 (1919): 359–82.

4. Guyatt, "Better Legs," 312.

5. For America's reputation as world leaders in artificial limb manufacture, see E. Muirhead Little, "Notes on Artificial Limbs for Sailors and Soldiers," *American Journal of Orthopedic Surgery* 15 (1917): 596–602. For examples of America's reputation in Germany, see Perry, "Re-Arming the Disabled Veteran."

6. Steven Mihm, "A Limb Which Shall Be Presentable in Polite Society," in Katherine Ott, David Serlin, and Stephen Mihm, eds., *Artificial Parts, Practical Lives: Modern Histories of Prosthetics* (New York: New York University Press, 2002): 282. For accounts and statistics of the degree to which America lagged behind European countries in preventing industrial accidents, see David Rosner and Gerald Markowitz, "The Early Movement for Occupational Safety and Health, 1900–1917," in Judith Walzer Leavitt and Ronald L. Numbers, eds., *Sickness and Health in America*, 3rd ed. (Madison: University of Wisconsin Press, 1997), esp. 468; Mark Aldrich, "Train Wrecks to Typhoid Fever: The Development of Railroad Medicine Organizations, 1850 to World War I," *Bulletin of the History of Medicine* 75, no. 2 (2001): 254–89; Gary Gerstle, *Working-Class Americanism: The Politics of Labor in a Textile City, 1914–1960,* 2nd ed. (Princeton: Princeton University Press, 2002); and David Brody, *Workers in Industrial America: Essays on the Twentieth-Century Struggle,* 2nd ed. (New York: Oxford University Press, 1993).

7. For number of artificial limb companies, see Katherine Ott, "The Sum of Its Parts: An Introduction to Modern Histories of Prosthetics," in Katherine Ott, David Serlin, and Stephen Mihm, eds., *Artificial Parts, Practical Lives: Modern Histories of Prosthetics* (New York: New York University Press, 2002): 26. For more on Roehampton, see Guyatt, "Better Legs."

8. *MDWW* 1: 429. The members of the Committee on Medicine included Army Surgeon General William Gorgas, Navy Surgeon General W. C. Braisted, US Public Health Service Surgeon General Rupert Blue, and civilian physicians W. J. Mayo, C. H. Mayo, and William H. Welch. The committee was responsible for developing a wide array of subcommittees ranging from the Committee on Standardization and the Committee on Venereal Disease to the Committee on Industrial Medicine and Surgery. For more on the Committee on Medicine, see *MDWW* 1: 559–65, and for references to the Committee on Standardization, see p. 559.

9. Ott makes a similar point about the relationship between surgeons and limb makers in "The Sum of Its Parts," 13.

10. Lisa Herschbach, "Prosthetic Reconstructions: Making the Industry, Re-Making the Body, Modelling the Nation," *History Workshop Journal* 44 (Autumn 1997): 22–57, 28.

11. For statistics, see *MDWW* 11: 71–72.

12. Edward Slavishak makes a similar point in his wonderful analysis of prosthetists in Pittsburg; see "Artificial Limbs and Industrial Workers' Bodies in Turn-of-the-Century Pittsburgh," *Journal of Social History* 37, no. 2 (2003): 369.

13. According to Herschbach, "the empty sleeve was . . . a badge of courage" during the postbellum years; see "Prosthetic Reconstructions," 24. Also, a 1903 issue of the *Atlanta Constitution* hailed the image of Johnny Reb—an aged, one-legged Confederate veteran hero, who tottered around with the

aid of a crutch, a peg leg and a cane—as a symbol of "supreme manliness," of the "moral perfection" found in a "devoted" soldier. See R. B. Rosenburg, " 'Empty Sleeves and Wooden Pegs': Disabled Confederate Veterans in Image and Reality," in David Gerber, ed., *Disabled Veterans in History* (Ann Arbor: University of Michigan Press, 2000): 204–28. See also Laurann Figg and Jane Farrell-Beck, "Amputation in the Civil War: Physical and Social Dimensions," *Journal of the History of Medicine and Allied Sciences* 48 (1993): 454–75.

14. David Silver to Army Surgeon General, 28 February 1918, NARA II, RG 112, box 309, file 442.3. For more about the national fears concerning disabled beggars taking to the streets, see Susan M. Schweik, *The Ugly Laws: Disability in Public* (New York: New York University, 2009). A 1910 US Immigration Commission survey demonstrated that Americans thought that "physical disability of a breadwinner" was one of the outstanding causes that led to economic dependency; Amy Fairchild, *Science at the Borders: Immigrant Medical Inspection and the Shaping of the Modern Industrial Labor Force* (Baltimore: Johns Hopkins University Press, 2003), 50.

15. David Silver to Army Surgeon General, 28 February 1918, NARA II, RG 112, box 309, file 442.3, and Herschbach, "Prosthetic Reconstructions."

16. For statistics of limb use in Canada, see J. A. C. Chandler, Chief of Division of Rehabilitation to Federal Board of Vocational Education District Vocational Officers, Washington, D.C. 19 February 1919, NARA II, RG 112, box 389, file 707.2. According to Chandler only 2 percent of Canadian soldiers supplied with replacement arms found them "satisfactory." The other 98 percent "had given up trying to use" their artificial limbs. A similar situation arose in Germany. See Perry, "Re-Arming the Disabled Veteran."

17. "Program of the Thirty-First Annual Meeting of the American Orthopedic Association," Pittsburgh, May 31–June 2, 1917, reprinted in the *American Journal of Orthopedic Surgery* 15 (1917): 426–37.

18. David Silver, "Lessons from the War in the Care of Industrial Injuries of the Extremities," *West Virginia Medical Journal* 14 (September 1919): 86.

19. "The Committee on Medicine, Council of National Defense," *MDWW* 1: 559.

20. David A. Hounshell, *From the American System to Mass Production, 1800–1932: The Development of Manufacturing Technology in the United States* (Baltimore: Johns Hopkins University Press, 1984), and Merritt Roe Smith, *Harpers Ferry Armory and the New Technology: The Challenge of Change* (Ithaca, NY: Cornell University Press, 1977).

21. Lisa Herschbach, "Prosthetic Reconstructions,"43.

22. According to Jennifer Keene, while the federal government conscripted soldiers during the Civil War, the Union Army ultimately only drafted 8 percent of its wartime force; see *Doughboys, the Great War, and the Remaking of America* (Baltimore: Johns Hopkins University Press, 2001). For another overview of U.S involvement in the Great War, see David M. Kennedy, *Over*

Here: The First World War and American Society (New York: Oxford University Press, 1980).

23. Keene, *Doughboys*, 2, 5, 20.
24. US Department of Agriculture Forest Service, "The Use of Wood In the Manufacture of Artificial Limbs," prepared for the Surgeon General's Office, War Department, 22 January 1918, NARA II, RG 112, box 309, file 442.3.
25. Ibid.
26. See Robert B. Osgood, Letterman General Hospital to the Office of the Surgeon General, Washington, D.C., 29 August 1918, and Robert L. Hull, Letterman General Hospital to the Office of the Surgeon General, Washington, D.C., September, 1918, NARA II, RG 112 "SGO 1917–1927, Letterman," box 101, file 721.1.
27. David Silver, "The Etiological Importance of Abnormal Foot Posture in Affections of the Knee," *American Journal of the Medical Sciences* 136 (1908): 726. See also David Silver to D. P. Christmann, Delta, Ohio, 3 December 1917, NARA II, RG 112, box 309, file 442.3. For secondary source treatment of the history of biomechanics, see Hillel Schwartz, "Torque: The New Kinaesthetic of the Twentieth Century," in Jonathan Crary and Sanford Kwinter, eds., *Incorporations* (New York: Zone Books, 1992), 71–126.
28. Ott, "The Sum of its Parts," 26.
29. Association of Artificial Limb Manufacturers (AALM) of the United States to David Silver, Washington, D.C., 19 October 1917, NARA II, RG 112, box 309, file 442.3.
30. For instance, when prosthetist Emory Staggs, who claimed to have "supplied most of the artificial limbs to the Government during the building of the Panama Canal," asked for a government contract to supply the War Department limbs for World War I, David Silver denied his request. Emory Staggs, Staggs Aluminum-Rawhide Artificial Limb Co., New York, NY to War Department, Washington, D.C., 5 November, 1917, NARA II, RG 112, box 309, file 442.3.
31. David Silver, Washington D.C. to Major Joel E. Goldthwait, AEF, France, 11 December 1917, NARA II, RG 112, box 309, file 442.3. For the Canadian system of artificial limb supply, see Clarence L. Starr, "The Role of Orthopedic Surgery in Modern Warfare," *American Journal of Orthopedic Surgery* 16 (1918): 421.
32. Slavishak, "Artificial Limbs and Industrial Workers' Bodies," 376. J. F. Rowley, Rowley Artificial Limb Co., Chicago, to David Silver, Washington, D.C., 3 December 1917, NARA II, RG 112, box 309, file 442.3.
33. Yet, as demonstrated in chap. 2, many orthopedic surgeons were disabled as well. For Rowley's work in England, see David Silver to Elliot Brackett, Washington, D.C., 18 October 1917, NARA II, RG 112, box 309, file 442.3.
34. J. F. Rowley, Rowley Artificial Limb Co., Chicago, to David Silver, Washington, D.C., 3 December 1917, NARA II, RG 112, box 309, file 442.3.
35. For Carnes's story, see Carnes Artificial Limb Company, *The Carnes Arm*

Puts You on the Pay Roll (Kansas City: Chas E. Brown Printing Co, 1913). For
Rowley, see Slavishak, "Artificial Limbs and Industrial Workers' Bodies."

36. Frederick M. Voss, employee of AA Marks, Geneva, Switzerland to David
Silver, Washington, D.C., 25 November 1917, NARA II, RG 112, box 309,
file 442.3. See also E. H. Erickson, Erickson Artificial Limb Co. to David
Silver, Washington, D.C., 19 October 1917, NARA II, RG 112, box 309, file
442.3. Erickson wrote that "in my 25 years experience as a wearer, maker
and fitter of limbs, I have learned many things relative to their mecha-
nisms and construction."

37. For an astute analysis of the showmanship qualities of artificial limb makers during the nineteenth century, see Herschbach, "Prosthetic Reconstructions."

38. E. H. Erickson, Erickson Artificial Limb Co. to David Silver, Washington,
D.C., 27 February 1917, NARA II, RG 112, box 309, file 442.3. See also
James G. Waltz , Merrick-Hopkins Co, Indianapolis, Ind., to David Silver,
Washington, D.C., 25 February 1918 NARA II, RG 112, box 309, file 442.3.

39. Doerflinger Artificial Limb Co, Milwaukee, Wisconsin to David Silver,
Washington D.C., October 1917, NARA II, RG 112, box 309, file 442.3, and
US Department of Agriculture Forest Service, "The Use of Wood In the
Manufacture of Artificial Limbs," prepared for the Surgeon General's Office,
War Department, 22 January 1918, NARA II, RG 112, box 309, file 442.3,
n.p.

40. J. E. Hanger (trade literature), *Solvutur Ambulando: A Symposium on Prosthetic
Achievement*, 1936, p. 4, quoted in Guyatt, "Better Legs," 314. Statements
such as these can be found across the trade. See, for instance, Doerflinger
Artificial Limb Co., Milwaukee, Wisconsin, to David Silver, Washington
D.C., October 1917, NARA II, RG 112, box 309, file 442.3, and James G.
Waltz , Merrick-Hopkins Co., Indianapolis, Ind., to David Silver, Washing-
ton, D.C., 25 February 1918 NARA II, RG 112, box 309, file 442.3.

41. David Silver to A. A. Marks, New York, New York, 19 February 1918, NARA
II, RG 112, box 308, file 442.3.

42. Doerflinger Artificial Limb Co., Milwaukee, Wisconsin, to David Silver,
Washington, D.C., October 1917, NARA II, RG 112, box 309, file 442.3.
For David Silver's views on the matter, see David Silver, Memo, 7 February
1918, Re: "Lack of Standardization of methods in Artificial limb construction," NARA II, RG 112, box 309, file 442.3.

43. David Silver to Major Elliot Brackett, Washington, D.C., 18 October 1917,
NARA II, RG 112, box 308, file 442.3.

44. With only a handful a limb companies involved in the use of fiber, David
Silver sent inquiries to several timber companies asking them if they would
be interested in participating in his investigation in the use of wood fiber.
For an example of such an inquiry, see David Silver to Mr. Ralph Lowry,
Beaver Board Company, Buffalo, NY, 11 October 1917, NARA II, RG 112,
box 309, file 442.3.

45. US Department of Agriculture Forest Service, "The Use of Wood in the Manufacture of Artificial Limbs," prepared for the Surgeon General's Office, War Department, 22 January 1918, NARA II, RG 112, box 309, file 442.3.
46. Albert G. Follett, President, E-Z-Fit Artificial Limb Co, New York, NY to David Silver, Washington, D.C., 14 December 1917, NARA II, RG 112, box 309, file 442.3.
47. US Department of Agriculture Forest Service, "The Use of Wood in the Manufacture of Artificial Limbs."
48. Albert G. Follett, President, E-Z-Fit Artificial Limb Co., New York, NY to David Silver, Washington, D.C., 14 December 1917, NARA II, RG 112, box 309, file 442.3.
49. David Silver to Major Joel E. Goldthwait, AEF, France, 11 December 1917, NARA II, RG 112, box 309, file 442.3.
50. David Silver, memorandum, "First Report on Amputation Cases," October 1918, NARA II, RG 112, box 431.
51. David Silver, "The Care of the Amputated in War," *Pennsylvania Medical Journal* 22 (January 1919): 213.
52. David Silver, memorandum, "First Report on Amputation Cases," October 1918, NARA II, RG 112, box 431.
53. Division of Military Orthopedic Surgery, Office of the Surgeon General, "Instruments and Appliances: Temporary Artificial Limbs," *Military Surgeon* 42 (1918): 490–98.
54. On the open market, E-Z-Legs retailed for approximately $50. David Silver, Memo, 4 December 1917, NARA II, RG 112, box 309, file 442.3. For costs of various limbs, including the E-Z-Leg, see David Silver, Memo, "Temporary Leg," n.d., and Dr. Albert H. Freiberg, Walter Reed General Hospital, memo, 18 July 1918. Both of these memos are found in NARA II, RG 112, box 309, file 442.3.
55. War Department, Office of the Surgeon General, "Information on Artificial Limbs and the Care of the Stump: A Manual for the Amputated," 1918, NARA II, RG 112, box 4608, file 443.
56. Albert G. Follett, President, E-Z-Fit Artificial Limb Co, New York, NY to David Silver, Washington, D.C., 14 December 1917, NARA II, RG 112, box 309, file 442.3.
57. "Amputation Service A.E.F.: Organization and Development," in *MDWW* 11: 703–4.
58. Ibid.
59. David Silver to Major Joel E. Goldthwait, AEF, France, 11 December 1917, NARA II, RG 112, box 309, file 442.3.
60. David Silver, "The Care of the Amputated in War," 213.
61. David Silver to Major Joel E. Goldthwait, AEF, France, 11 December 1917, NARA II, RG 112, box 309, file 442.3.
62. For more on disability and the media, see Paul Longmore, "Screening Stereotypes: Images of Disabled People in Television and Motion Pictures," in

Why I Burned My Book and Other Essays on Disability (Philadelphia : Temple University Press, 2003): 131–47. See also Martin F. Norden, *The Cinema of Isolation: A History of Physical Disability in the Movies* (New Brunswick, NJ: Rutgers University Press, 1994) and Sharon Snyder and David T. Mitchell, *Cultural Locations of Disability* (Chicago: University of Chicago Press, 2006).

63. Bill Brown, "Science Fiction, the World's Fair, and the Prosthetics of Empire, 1910–1915," in Amy Kaplan and Donald E. Pease, eds., *Cultures of United States Imperialism* (Durham: Duke University Press, 1993): 135. Also, while "The Thieving Hand" played in the local film houses, J. M. Barrie's "Peter Pan," with the unsavory one-armed "Captain Hook," ran on Broadway from 1905 to 1914 and was eventually filmed in 1924.

64. "New Artificial Leg Gives Sure Footing," *New York Times*, March 22, 1918, A11.

65. David Silver to Army Surgeon General, 28 February 1918, NARA II, RG 112, box 309, file 442.3.

66. Perry, "Re-Arming the Disabled Veteran." For Great Britain, see Jeffrey Reznick, *Healing the Nation: Soldiers and the Culture of Caregiving in Britain during the Great War* (Manchester: Manchester University Press, 2004).

67. David Silver to Army Surgeon General, 28 February 1918, NARA II, RG 112, box 309, file 442.3.

68. David Silver, Washington, D.C. to Major Joel E. Goldthwait, AEF, France, 11 December 1917, NARA II, RG 112, box 309, file 442.3.

69. Perry, "Re-Arming the Disabled Veteran," 96.

70. W. S. Dobbs, Advisor on Amputation Cases for the Director of Vocational Training to the Office of the Surgeon General, Ottawa, Canada, 6 October 1919, NARA II, RG 112, box 306, file 442.3.

71. David Silver, Washington, D.C., to Major Joel E. Goldthwait, AEF, France, 11 December 1917, NARA II, RG 112, box 309, file 442.3.

72. *MDWW* 11: 746.

73. Report found in David Silver, memorandum, "First Report on Amputation Cases," October 1918, NARA II, RG 112, box 431.

74. Chief Medical Advisor, Bureau of War Risk Insurance, to US Surgeon General, Washington, D.C., 27 May 1919, NARA II, RG 112, box 309, file 442.3.

75. War Department, Office of the Surgeon General, "Information on Artificial Limbs and the Care of the Stump: A Manual for the Amputated," 1918, NARA II, RG 112, box 4608, file 443.

76. Lt. Col. MacLeod, Commandant of Queen Mary's Hospital Roehampton, *Daily Sketch*, October 1919, n.p., as quoted in Guyatt, "Better Legs," 314.

77. The subject of race, ethnicity, and disability has received relatively little scholarly attention. Some notable exceptions include Susan Burch and Hannah Joyer, *Unspeakable: The Story of Junius Wilson* (Chapel Hill: University of North Carolina Press, 2007); Ellen Dwyer, "Psychiatry and Race during World War II," *Journal of the History of Medicine and Allied Sciences* 61, no. 2 (2006): 117; and Jennifer D. Keene, "Protest and Disability: A New

Look at African American Soldiers during the First World War," in Pierre Purseigle, ed., *Warfare and Belligerence: Perspectives in First World War Studies* (Leiden: Brill, 2005), 215–41.

78. Slavishak makes a similar point about limb makers and how they "modeled their products after the appearance of Anglo-American body parts"; see "Artificial Limbs and Industrial Workers' Bodies" 373.

79. US Department of Agriculture Forest Service, "The Use of Wood in the Manufacture of Artificial Limbs," prepared for the Surgeon General's Office, War Department, 22 January 1918, NARA II, RG 112, box 309, file 442.3, n.p.

80. Leo Mayer, "The Organization and Aims of the Orthopedic Reconstruction Hospital," *American Journal of the Care for Cripples* 5 (September 1917): 81.

81. The rigid steel construction of the hook-arm, according to David Silver, proved to be more practical and useful in completing everyday tasks, especially for "farmers, plumbers, machinists . . . and men of that class." See David Silver to Major Joel E. Goldthwait, AEF, France, 11 December 1917, NARA II, RG 112, box 309, file 442.3. See also the catalog for the Dorrance Utility Arm, Chicago Artificial Limb Co., San Francisco, Ca., Dr. I. R. Fenner, Manager, NARA II, RG 112, box 309, file 442.3.

82. Perry, "Re-Arming the Disabled Veteran," 97.

83. Silver, "The Care of the Amputated in War," 214.

84. Roy Rosenzweig, *Eight Hours for What We Will: Workers and Leisure in an Industrial City, 1870–1920* (Cambridge: Cambridge University Press, 1983), and Gail Bederman, *Manliness and Civilization: A Cultural History of Gender and Race in the United States, 1880–1917* (Chicago: University of Chicago Press, 1995).

85. The estimates of total amputees from World War I are as follows: loss of upper extremity: 2,346; loss of lower extremity: 2,032; upper and lower extremities: 25; total: 4403. See *MDWW* 11: 713–48.

86. Ibid.

87. By World War II, the military issued legs with hydraulic knees and plastic sockets. Contrary to the Great War, the driving philosophy of World War II was that limbs must be created to fit each individual patient. See National Research Council, *Human Limbs and Their Substitutes, Presenting Results of Engineering and Medical Studies of the Human Extremities and Application of the Data to the Design and Fitting of Artificial Limbs and to the Care and Training of Amputees* (New York: McGraw-Hill, 1954). For more on disability and prosthetics during the era of World War II, see David Serlin, *Replaceable You: Engineering the Body in Postwar America* (Chicago: University of Chicago Press, 2004).

CHAPTER SIX

1. For a sampling of the literature on Wilson and the Espionage Act, see David Kennedy, *Over Here: The First World War and American Society* (New

York: Oxford University Press, 1980), 25–66; Robert Zieger, *America's Great War: World War I and the American Experience* (New York: Rowman and Littlefield, 2000), 78–81; John A. Thompson, *Reformers and War: American Progressive Publicists and the First World War* (Cambridge: Cambridge University Press, 1987); John Braeman, "World War One and the Crisis of American Liberty," *American Quarterly* 15 (Spring 1964): 104–12; Donald Johnson, "Wilson, Burleson, and Censorship in the First World War," *Journal of Southern History* 28 (February 1962): 46–58.

2. Ana Carden-Coyne, "Ungrateful Bodies: Rehabilitation, Resistance, and Disabled American Veterans of the First World War," *European Review of History* 14, no. 4 (2007): 543–65.

3. See interviews with *Carry On* editor, Maj. Casey Wood, and the director of Army Physical Reconstruction, Lt. Col. Frank Billings in Frank Parker Stockbridge, "Putting Our War Cripples Back on the Payroll," *New York Times*, 12 May 1918, 80.

4. Judge Julian W. Mack, "A Chance—With a Running Start," *Carry On* 1, no. 2 (August 1918): 13.

5. Dr. Casey A. Wood, chief editor of *Carry On*, stated that he was attempting to make the "whole community assume a proper attitude towards the returned invalided soldier." See Casey A. Wood to Mr. T. B. Kidner, Federal Board for Vocational Education, Washington, D.C., 27 August 1918, NARA I, RG 15, box 254, file 1.

6. Gelett Burgess, "Victim versus Victor," *Carry On* 1, no. 1 (July 1918): 20–22, 22.

7. K. Walter Hickel, "Entitling Citizens: World War I, Progressivism, and the Origins of the American Welfare State, 1917–1928" (PhD diss., Columbia University, 1999), 177. Liability laws made employers greatly reluctant to hire men and women whose range and control of movement were limited, and who might therefore be at a higher risk of suffering injury in the workplace for which employers could be held legally responsible under these laws.

8. Casey Wood, "Carry On," *American Journal of Orthopedic Surgery* 16, no. 8 (August 1918): 539.

9. For more on the history of this branch of the US Army, see Joseph W. A. Whitehorne, *The Inspectors General of the United States Army, 1903–1939* (Washington, DC: Office of the Inspector General and Center of Military History, United States Army, 1998).

10. Jennifer Keene discusses the risks in her essay about black soldiers and their letters of complaints sent to Emmett J. Scott, secretary of the Tuskegee Institute, who served as a special assistant to the secretary of war during World War I; see Keene, "Protest and Disability: A New Look at African American Soldiers during the First World War," in Pierre Purseigle, ed., *Warfare and Belligerence: Perspectives in First World War Studies* (Leiden: Brill Press, 2005), 215–41.

11. *MDWW* 5: 360. The constitution of the Fifty-Fifty League stated that "Our purpose is to cooperate with those in authority in carrying out law and order and make it possible for them to grant a maximum of liberty for all concerned. We believe that by force of public opinion we can persuade men to so conduct themselves that a higher degree of law and order will result, thus rendering many of the present regulations and restriction unnecessary." Another one of the league's goals was to "foster and perpetuate real Americanism."

12. Sworn testimony of Cadet William Dearing Davis, NARA II, RG 159, box 1109, "Office of the Inspector General Correspondence, 1917–1934, Walter Reed," folder 4.

13. This same complaint can be found in Lena Hitchcock, "The Great Adventure" (unpublished memoir), vii.

14. For Jackson's patient record, see sworn testimony of Cpl. Leonard E. Jackson, NARA II, RG 159, box 1109, "Office of the Inspector General Correspondence, 1917–1934, Walter Reed," folder 4.

15. In other words, army officials delegitimized Davis's protests by making him out to be a hysteric of sorts, a character flaw that stood in opposition to the rehabilitation officials' notion of manliness. For more on the history of hysteria and manliness, see Paul Frederick Lerner, *Hysterical Men: War, Psychiatry, and the Politics of Trauma in Germany, 1890–1930* (Ithaca, NY: Cornell University Press, 2003), and Mark S. Micale, *Hysterical Men: The Hidden History of Male Nervous Illness* (Cambridge, MA: Harvard University Press, 2008).

16. For Truby's comments, see sworn testimony of Col. Willard F. Truby, NARA II, RG 159, box 1109, "Office of the Inspector General Correspondence, 1917–1934, Walter Reed," folder 4.

17. Lt. Col. Harry E. Mock, "Physical Training," *Carry On* 1, no. 8 (May 1919): 5–8.

18. W. Frank Persons, "Looking After the Soldier's Family," *Carry On* 1, no. 1 (June 1918): 29–31.

19. Gail Bederman, *Manliness and Civilization: A Cultural History of Gender and Race in the United States, 1880–1917* (Chicago: University of Chicago Press, 1995).

20. For a description of this ranking among disabled soldiers, see Frank Ward O'Mally, "Home Comers from France," *Carry On* 1 no. 3 (September 1918): 5–9.

21. General John Pershing to Major General Henry T. Allen, 90th Division, October 24, 1918, "1918 (Jan-Sept.)" folder, box 11, Henry T. Allen Papers, Library of Congress, quoted in Jennifer Keene, *Doughboys, the Great War, and the Remaking of America* (Baltimore: Johns Hopkins University Press, 2001), 65.

22. These numbers come from Keene, *Doughboys*, 5.

23. Ibid., 36, 57.

24. Sworn testimony of Jerome E. Lane, NARA II, RG 159, box 845, "Office of the Inspector General Correspondence, 1917–1934, Letterman General Hospital," folder 2.

25. Sworn testimony of Private John S. Hall, NARA II, RG 159, box 1109, "Office of the Inspector General Correspondence, 1917–1934, Walter Reed," folder 3.

26. Private John Hall, Georgia to Mrs. Hall (mother), Scranton, PA, 7 October 1918, NARA II, RG 159, box 1109, "Office of the Inspector General Correspondence, 1917–1934, Walter Reed," folder 3.

27. For more on malingering, see Roger Cooter, "Malingering in Modernity: Psychological Scripts and Adversarial Encounters during the First World War," in Roger Cooter, Mark Harrison, and Steve Sturdy, eds., *War, Medicine and Modernity* (Stroud: Sutton, 1998), 125–48.

28. For an overview and interpretation on the virtues of manhood and masculinity at the time, see Bederman, *Manliness and Civilization*. For the best survey on the history of physical culture and novelistic literature portraying manliness at the time, see John F. Kasson, *Houdini, Tarzan, and the Perfect Man: The White Male Body and the Challenge of Modernity in America* (New York: Hill and Wang, 2001). For an account of the importance of war on masculinity, see Christina S. Jarvis, *The Male Body at War: American Masculinity during World War II* (DeKalb: Northern Illinois University Press, 2004), and Susan Jeffords, *The Remasculinization of America: Gender and the Vietnam War* (Bloomington: Indiana University Press, 1989). More recently, disability historians have begun to study how gendered assumptions inform societal definitions of the disabled body. One of the best treatments of gender and disability, especially of maimed soldiers and veterans, is David A. Gerber, ed., *Disabled Veterans in History* (Ann Arbor: University of Michigan Press, 2000; see, more specifically, his "Introduction: Finding Disabled Veterans in History," 1–51.

29. Herbert Kaufman, "Not Charity—But a Chance," *Carry On* 1 (June 1918): 11–12, 12, and Charles M. Schwab, "Launching Men Anew: The Seas of Opportunity Are Waiting for Specialized Brains," *Carry On* 1 (August 1918): 6–8, 6.

30. Burgess, "Victim versus Victor."

31. Harry E. Mock, "The Way Out: Desire and Ambition Must be Born in the Man Himself," *Carry On* 1, no. 2 (August 1918): 27.

32. Rupert Hughes, "The Lucky Handicap," *Carry On* 1, no. 3 (September 1918): 11–12.

33. Disability studies scholars define "supercrips" as people who go to great lengths to overcome their disabilities. Supercrips often receive great praise among the ablebodied. Some well-known examples of "supercrips" include Helen Keller and Franklin Delano Roosevelt. For more, see Kim E. Nielsen, *The Radical Lives of Helen Keller* (New York: New York University Press,

2004); and Hugh G. Gallagher, *FDR's Splendid Deception* (New York: Dodd, Mead, 1985).

34. "From His Neck Up," *Carry On* 1, no. 1 (June 1918): 23.

35. Maury Maverick, *A Maverick American* (New York: Convici Friede, 1937).

36. See Richard B. Henderson, *Maury Maverick: A Political Biography* (Austin: University of Texas Press, 1970), and Maverick, *A Maverick American.*

37. Miss Louise Irving Capen, Northhampton, Massachusetts to George Creel, Washington, D.C., 24 July 1918, NARA II, RG 159, box 1109, "Office of the Inspector General Correspondence, 1917–1934, Walter Reed," folder 3.

38. See Inspector General reports from both Walter Reed and Letterman Hospitals. For Walter Reed, see NARA II, RG 159, box 1109. For Letterman, see NARA II, RG 159, box 845.

39. Col. Truby, as reported by Mrs. B. R. Russell in sworn testimony of Mrs. B. R. Russell, NARA II, RG 159, box 1109, "Office of the Inspector General Correspondence, 1917–1934, Walter Reed," folder 6. For more on the negative portrayal of meddlesome women, see "When a Feller Needs a Friend," *Carry On* 1, no. 1 (June 1918): 19.

40. Major General Merritte W. Ireland, "Carry On," *Carry On* 1, no. 6 (March 1919): 4.

41. Douglas McMurtrie, *The Organization, Work and Method of the Red Cross Institute for Crippled and Disabled Men* (New York: Red Cross Institute for Crippled and Disabled Men, 1918), 21.

42. "No Wonder Buddy Don't Care to Mend," *The Come-Back,* April 2, 1919, 8.

43. This is not to say that all World War I disabled soldiers were heterosexual, but sexual orientation is difficult to determine, especially in a context heavily dominated by heterosexual norms. As in other wars of the twentieth century, World War I disabled soldiers frequently engaged in cross-dressing as a form of entertainment; see Army Medical Museum Photographs, Otis Historical Archives, National Museum of Health and Medicine, Washington, DC This kind of behavior has been best analyzed by David Serlin, "Crippling Masculinity: Queerness and Disability in US Military Culture, 1800–1945," *GLQ: Journal of Lesbian and Gay Studies* 9 (2003): 149–79.

44. "Surely, Girlies Get Tired, Too," *The Come-back,* April 9, 1919, 7.

45. "Cripple Shimmy Latest at Reed: Birds Shy on Legs Shake Brutal Feet at Red Cross Hop," *The Come-Back,* April 9, 1919, 2.

46. B. Mateson, "That Sweater From the ONLY Girl," *The Come-Back,* May 14, 1919, 6, emphasis in original.

47. Emmett J. Scott, *The American Negro in the World War,* 2nd ed. (New York: Arno Press, 1969), 447. Scott's original history of black involvement in World War I was first published in 1919.

48. For more on African Americans fighting for the French, see Keene, *Doughboys,* and Adriane Lentz-Smith, *Freedom Struggles: African Americans and World War I* (Cambridge, MA: Harvard University Press, 2009).

49. C. R. Darnall, Washington, D.C. to Emmett J. Scott, Special Assistant to Secretary of War, War and Navy Building, Washington, D.C., 19 December 1918, NARA I, RG 15, box 254, "VA Rehabilitation Division, General Correspondence, 1918–1925," folder "December 1918."
50. Keene, *Doughboys*, 23.
51. Keene, "Protest and Disability," 224.
52. William Lloyd Imes, Young Men's Christian Association (YMCA) secretary, to Emmett J. Scott, August 18, 1918, file #10218-209, Textual Records of the War Department General and Special Staffs, NARA II, 1918, RG 165, as quoted in Keene, *Doughboys*, 95.
53. Diary of Private Guy R. Moore, "World War I Veterans Survey, Medical Hospital, Base Hospital, 80-up," Carlisle Barracks, Archives and Manuscripts, Carlisle, Pennsylvania.
54. Sworn testimony of Jerome E. Lane, NARA II, RG 159, box 845, "Office of the Inspector General Correspondence, 1917–1934, Letterman General Hospital," folder 2.
55. Private James Cunningham, Walter Reed Hospital to Mrs. R. B. Russell, 7 July 1918, NARA II, RG 159, box 1109, "Office of the Inspector General Correspondence, 1917–1934, Walter Reed," folder 6.
56. Sworn testimony of Mrs. B. R. Russell, NARA II, RG 159, box 1109, "Office of the Inspector General Correspondence, 1917–1934, Walter Reed," folder 6.
57. Joel Moore to the National Association for the Advancement of Colored People (NAACP), December 29, 1923, "Military, Gen'l, 1924, Jan-Dec" folder, box C-375, series 1, NAACP papers, Library of Congress, Washington, DC. As quoted in Keene, "Protest and Disability," 235.
58. Vanessa Northington Gamble, *Making a Place for Ourselves: The Black Hospital Movement, 1920–1945* (New York: Oxford University Press, 1995), 73.
59. Hickel estimates that only 3.5 percent of all postwar vocational trainees were black; see "Entitling Citizens," 258.
60. See Edward D. Berkowitz, "The American Disability System in Historical Perspective," in Edward D. Berkowitz, ed., *Disability Policies and Government Programs* (New York: Praeger, 1979): 45.
61. Hickel, "Entitling Citizens," 255.
62. Paul Dickson and Thomas B. Allen, *The Bonus Army: An American Epic* (New York: Walker & Co, 2005).
63. Hickel, "Entitling Citizens," 174.
64. R. J. Fuller, "The United States Veterans Bureau: Progress and Outlook," March 31, 1924, pp. 12–21A, 8, file 10.5, "Addresses," box 1, Office File of Earnest B. Luce, 1919–1923, NARA I, RG 15. As quoted in Hickel, "Entitling Citizens," 204.
65. For more on Cassiday's intriguing story, see Dickson and Allen, *The Bonus Army*.
66. Sworn testimony of 1st Lt. Philip Nelson, 27 July 1920, NARA II, RG 159,

box 1109, "Office of the Inspector General Correspondence, 1917–1934, Walter Reed," folder 13.

67. Douglas McMurtrie, "The High Road to Self-Support," *Carry On* 1 (July 1918): 4.

68. "The Enemy Was Ready: How Germany Made Preparation for Her Wounded," *Carry On* 1, no. 1 (June 1918): 24–28.

69. Samuel Harden Church, "So Much for So Much: Give the Returned Soldier a Job and Pay Him What he Earns," *Carry On* 1, no. 4 (October–November, 1918): 9–10.

70. Martin Pernick, *The Black Stork: Eugenics and the Death of "Defective" Babies in American Medicine and Motion Pictures Since 1915* (New York: Oxford University Press, 1996).

71. Daniel J. Kevles, *In the Name of Eugenics: Genetics and the Uses of Human Heredity* (New York: Knopf, 1985), 91–92.

72. For a more thorough discussion of the prevalence of vocational training in Progressive Era America, see Hickel, "Entitling Citizens," 171–202; percentages can be found on p. 200.

73. Hughes, "The Lucky Handicap," 11–12.

74. McMurtrie, "The High Road to Self Support," 7.

75. "From His Neck Up," *Carry On* 1, no. 1 (June 1918): 23. See also Charles M. Schwab, "Launching Men Anew: The Seas of Opportunity are Waiting for Specialized Brains," *Carry On* 1, no. 2 (August 1918): 6–8, 6.

76. Sworn testimony of Pvt. George Russo, 27 July 1920, NARA II, RG 159, box 1109, "Office of the Inspector General Correspondence, 1917–1934, Walter Reed," folder 13.

77. Sworn testimony of 1st Lt. Philip Nelson, 27 July 1920, NARA II, RG 159, box 1109, "Office of the Inspector General Correspondence, 1917–1934, Walter Reed," folder 13.

78. Sworn testimony of Pvt. Jeremiah Hurley, 27 July 1920, NARA II, RG 159, box 1109, "Office of the Inspector General Correspondence, 1917–1934, Walter Reed," folder 13.

79. For "parasites," see Burgess, "Victim versus Victor," 20.

80. Sarah Rose makes the poignant argument that a sizeable portion of the working class in the early twentieth century suffered from physical disabilities—that being able-bodied was the exception rather than the rule; see "'Crippled' Hands: Disability in Labor and Working-Class History," *Labor: Studies in Working-Class History of the Americas* 2 (2005): 27–54.

81. Clinical Record of Lee F. Steinbacker, NARA II, RG 159, box 845, "Office of the Inspector General Correspondence, 1917–1934, Letterman General Hospital," folder 5.

82. Marion Girard, *A Strange and Formidable Weapon: British Responses to World War I Poison Gas* (Lincoln: University of Nebraska Press, 2008).

83. It is also important to note Webster's play on common war verbiage and

imagery in this illustration. "To a Man's Land" resembles the term, "No Man's Land," a popular expression used to define the territory between the German and French trench systems.

84. Colonel William Thompson, "The Vision of a Veteran of the Sixties," *Carry On* 1, no. 7 (April 1919): 13–14, 13.

85. Sworn testimony of 1st Lt. Harold H. Tittman, 27 July 1920, NARA II, RG 159, box 1109, "Office of the Inspector General Correspondence, 1917–1934, Walter Reed," folder 13.

86. Maverick, *A Maverick American,* 135.

87. For more about the 1932 march, see Dickson and Allen, *The Bonus Army,* as well as Stephen R. Ortiz, *Beyond the Bonus March and GI Bill: How Veteran Politics Shaped the New Deal Era* (New York: New York University Press, 2010).

CHAPTER SEVEN

1. Edward D. Berkowitz, "The American Disability System in Historical Perspective," in Edward D. Berkowitz, ed., *Disability Policies and Government Programs* (New York: Praeger, 1979), 43.

2. David Silver, "Abstract on Orthopedic Surgery and the War," to be delivered at Medical Society, County of Kings, Brooklyn, NY, November 1917, NARA II, RG 112, SGO 730, box 431, folder 8.

3. See Sanders Marble, *Rehabilitating the Wounded: Historical Perspective on Army Policy* (Washington, DC: Office of Medical History, 2008), 14.

4. C. Esco Oberman, *A History of Vocational Rehabilitation in America* (Minneapolis: T. S. Denison and Company, Inc., 1965). After removing a clause granting rehabilitation to disabled civilian workers, Congress passed the FVRA in July 1918.

5. For various interpretations about why the United States failed to create national health care insurance during the World War I era, see Beatrix Hoffman, *The Wages of Sickness: The Politics of Health Insurance in Progressive America* (Chapel Hill: University of North Carolina Press, 2001); Jonathan Engel, *Doctors and Reformers: Discussion and Debate over Health Policy, 1925–1950* (Columbia: University of South Carolina Press, 2002); Ronald L. Numbers, *Almost Persuaded: American Physicians and Compulsory Health Insurance, 1912–1920* (Baltimore: Johns Hopkins University Press, 1978); Colin Gordon, *Dead on Arrival: The Politics of Health Care in Twentieth Century America* (Princeton: Princeton University Press, 2003); Alan Derickson, *Health Security for All: Dreams of Universal Health Care in America* (Baltimore: Johns Hopkins University Press, 2005); and Daniel Fox, *Power and Illness: The Failure and Future of American Health Policy* (Berkeley: University of California Press, 1993).

6. See Numbers, *Almost Persuaded,* esp. chap. 9.

7. Edward Berkowitz makes this same point in *Disabled Policy: America's Pro-*

grams for the Handicapped (Cambridge: Cambridge University Press, 1987), 164.

8. Mark Aldrich, *Safety First: Technology, Labor, and Business in the Building of American Work Safety, 1870–1939* (Baltimore: Johns Hopkins University Press, 1997), 79.

9. Berkowitz, "The American Disability System in Historical Perspective," 16–74. Berkowitz demonstrates that business supported worker's compensation acts because they wanted protection against the mounting (and very expensive) lawsuits brought against them by injured employees.

10. David Rosner and Gerald Markowitz, "The Early Movement of Occupational Safety and Health, 1900–1917," in Judith Walzer Leavitt and Ronald L. Numbers, eds., *Sickness and Health in America*, 3rd ed. (Madison: University of Wisconsin Press, 1997), 479.

11. Hoffman, *The Wages of Sickness*, 48.

12. I. M. Rubinow, *Social Insurance* (New York: Arno, 1969).

13. Hoffman, *The Wages of Sickness*, 55.

14. Ibid., esp. chap. 8.

15. Harvey Kantor and David B. Tyack, "Historical Perspectives on Vocationalism in American Education," in Kantor and Tyack, eds., *Work, Youth, and Schooling: Historical Perspectives on Vocationalism in American Education* (Stanford: Stanford University Press, 1982), 2.

16. Harvey Kantor, "Vocationalism in American Education: The Economic and Political Context," in Kantor and Tyack, eds., *Work, Youth, and Schooling*, 26.

17. John Dewey, *Democracy and Education: An Introduction to the Philosophy of Education* (New York: Macmillan Co., 1916), 374.

18. See Howard R. D. Gordon, *The History and Growth of Career and Technical Education in America*, 3rd ed. (Long Grove, IL: Waveland Press, 2007), 29–32. See also, Kantor, "Vocationalism in American Education," 32.

19. Kantor and Tyack, "Historical perspectives on Vocationalism in American Education," 4.

20. In addition to the nation's major businesses, the roster of NSPIE supporters and participants included Jane Addams and AF of L president, Samuel Gompers. See Crystal Leigh Dunlevy, "The Contributions of Women of the National Society for the Promotion of Industrial Education to the Development of Vocational Education for Women, 1906–1917" (PhD diss., Rutgers University, 1988).

21. Arthur F. McClure, James Riley Chrisman, and Perry Mock, eds., *Education for Work: The Historical Evolution of Vocational and Distributive Education in America* (Rutherford, NJ: Fairleigh Dickinson University Press, 1985), 36–37.

22. Harvey Kantor, "Work, Education, and Vocational Reform: The Ideological Origins of Vocational Education, 1890–1920," *American Journal of Education* 94, no. 4 (August 1986): 416.

23. McClure, Chrisman, and Mock, *Education for Work*, 57.

24. For more on Smith, see Dewey W. Grantham, Jr., *Hoke Smith and the Politics of the New South* (Baton Rouge: Louisiana State University Press, 1958). See also Dewey W. Grantham, Jr., *The Regional Imagination: The South and Recent American History* (Nashville: Vanderbilt University Press, 1979), esp. chap. 8.

25. Karl Walter Hickel, "Entitling Citizens: World War I, Progressivism, and the Origins of the American Welfare State, 1917–1928" (PhD diss., Columbia University, 1999), 254. See also Howard R. D. Gordon, *The History and Growth of Career and Technical Education*. As Kantor rightly points out, no greater gap between rhetoric and reality exists than in the history of black vocational education. "Few groups have placed greater faith in the equalizing power of schooling yet received such meager economic returns for their additional education." Kantor and Tyack, "Historical Perspectives on Vocationalism in American Education," 10.

26. The bill appropriated $600,000 a year for agricultural extension programs, with funds allotted on the basis of the population of a state and its proportion of rural inhabitants.

27. For more on the role of US noncombatants in World War I, see Jennifer Keene, *Doughboys, the Great War, and the Remaking of America* (Baltimore: Johns Hopkins University Press, 2001), 50ff.

28. McClure, Chrisman, and Mock, *Education for Work*, 59–63.

29. For more details about the FBVE, see Stull W. Holt, *The Federal Board for Vocational Education: Its History, Activities, and Organization* (New York: Appleton, 1922); Richard Scotch, *From Good Will to Civil Rights*, 2nd ed. (Philadelphia: Temple University Press, 2001); and Larry Cuban "Enduring Resiliency: Enacting and Implementing Federal Vocational Education Legislation," in Kantor and Tyack, eds., *Work, Youth, and Schooling*, 45–78.

30. For numbers, see Howard R. D. Gordon, *The History and Growth of Career and Technical Education*, 66.

31. Senate Committee on Education and Labor, *Rehabilitation and Vocational Reeducation of Crippled Soldiers and Sailors*, 65th Cong., 2nd sess., February 1918, S. Doc. 173, 53.

32. Glenn Gritzer and Arnold Arluke, *The Making of Rehabilitation: A Political Economy of Medical Specialization, 1890–1980* (Berkeley: University of California Press, 1985), esp. chap. 3.

33. Senate Committee on Education and Labor, *Rehabilitation and Vocational Reeducation of Crippled Soldiers and Sailors*, S. Doc. 173, 36.

34. Senate Committee on Education and Labor, *Rehabilitation and Vocational Reeducation of Crippled Soldiers and Sailors*, S. Doc.173, 26. See also "Science to Rebuild Our War Cripples: Government Selects Sites for 19 Great 'Reconstruction' Hospitals in as Many Cities; Every Wounded Man to be the Nation's Ward until He is Entirely Fit for Civil Life," *New York Times*, 17 September 1917, A6.

35. Senate Committee on Education and Labor, *Rehabilitation and Vocational Reeducation of Crippled Soldiers and Sailors*, S. Doc. 173, 9.

36. Rosemary A. Stevens, *In Sickness and in Wealth: American Hospitals in the Twentieth Century*, 2nd ed. (Baltimore: Johns Hopkins University Press, 1999), 126.

37. *MDWW*, vol. 13, pt. 1 (Washington, DC: GPO, 1927), 9–10.

38. See Edwin F. Hirsch, *Frank Billings: A Leader in Chicago Medicine* (Chicago: University of Chicago, 1966).

39. Sworn testimony of Major John W. Shiels, NARA II, RG 159, box 845, "Office of the Inspector General Correspondence, 1917–1934, Letterman General Hospital," folder 5.

40. See, for example, Fred H. Albee, *A Surgeon's Fight to Rebuild Men* (New York: Dutton, 1943), 219.

41. Harry E. Mock, "The Disabled in Industry," *Annals of the American Academy of Political and Social Science* 80 (November 1918): 31. In order to advance the cause of both military and industrial medicine, Mock began holding night classes at Rush Medical College.

42. Statistics found in Roger Cooter, *Surgery and Society in Peace and War: Orthopaedics and the Organization of Modern Medicine, 1880–1948* (London: Macmillan Press, 1993), 140.

43. *MDWW*, vol. 15, pt. 2 (Washington, DC: GPO, 1925), 1018.

44. "New Artificial Leg Gives Sure Footing," *New York Times*, 22 March 1918, B11.

45. See Greg Eghigian, *Making Security Social: Disability, Insurance, and the Birth of the Social Entitlement State in Germany* (Ann Arbor: University of Michigan Press, 2000). For more on accident hospitals in Germany, see Thomas Schlich, "Trauma Surgery and Traffic Policy in Germany in the 1930s: A Case Study in the Coevolution of Modern Surgery and Society," *Bulletin of the History of Medicine* 80, no. 1 (2006): 73–94. For Great Britain, see Cooter, *Surgery and Society*, esp. chap. 7. Cooter maintains, however, that the National Health Insurance Act did not cover specialty services (p. 146). For a comparison of Germany, Great Britain, and the United States, see Deborah Stone, *The Disabled State* (Philadelphia: Temple University Press, 1984).

46. Mark Aldrich, "Train Wrecks to Typhoid Fever: The Development of Railroad Medicine Organizations, 1850 to World War I," *Bulletin of the History of Medicine* 75, no. 2 (2001): 254–89.

47. For a summary of the conference and its participants, see Senate Committee on Education and Labor, *Rehabilitation and Vocational Reeducation of Crippled Soldiers and Sailors*, S. Doc. 173, 29.

48. For contents of the OSG's tentative bill, see ibid., 26–46.

49. Gustavus Adolphus Weber, *The Employees' Compensation Commission: Its History, Activities, and Organizations* (New York and London: D. Appleton and Company, 1922).

50. House Committee on Education and Senate Committee on Education and Labor, *Vocational Rehabilitation of Disabled Soldiers and Sailors: Hearings on*

S. *4284 and H.R. 11367*, 65th Cong., 2nd sess., 30 April–2 May, 1918, 21. [Hereafter cited as *Hearings on S. 4284 and H.R. 11367*.]

51. Ibid., 88.

52. Senate Committee on Education and Labor, *Rehabilitation and Vocational Reeducation of Crippled Soldiers and Sailors*, S. Doc. 173, 61–68. McMurtrie submitted documentation for the record.

53. *Hearings on S. 4284 and H.R. 11367*, 25.

54. Senate Committee on Education and Labor, *Vocational Rehabilitation of Disabled Soldiers and Sailors*, 65th Cong., 2nd sess., January 1918, S. Doc. 166, 12.

55. Federal Vocational Rehabilitation Act of 1918, HR 11367, 65th Cong., 2nd sess., *Congressional Record* 56 (May 10–June 4, 1918): 6301–7382. For the surgeon general's definition, see p. 6957; see also pp. 7070, 7072.

56. *Hearings on S. 4284 and H.R. 11367*, 18, 35.

57. Ibid., 62, 19.

58. Ibid., 14.

59. *Congressional Record* 56: 6953 and 6955.

60. *Hearings on S. 4284 and H.R. 11367*, 77, 72.

61. Ibid., 71.

62. *Congressional Record* 56: 6963.

63. Spending during the 65th Congress was unprecedented. Historian Richard L. Watson estimates that between 1916 and 1919 annual federal expenditures increased 2,454 percent—a far greater percentage increase than took place during any other war, including World War II. See Watson, "A Testing Time for Southern Congressional Leadership: The War Crisis of 1917–1918," *Journal of Southern History* 44, no. 1 (February 1978): 3–40.

64. *Congressional Record* 56: 6964–65.

65. For "weakening of the party system," see Watson, "A Testing Time," 35.

66. For Ransdell's comments, see *Congressional Record* 56: 6958–59. Ransdell's position on the matter may be partially explained by the fact that he served on the Committee on Public Health and National Quarantine—a committee in charge of a great portion of the nation's federally funded hospitals in existence at the time—from the 63rd Congress to the 65th.

67. *Hearings on S. 4284 and H.R. 11367*, 72–73.

68. For Brandegee's remarks, see *Congressional Record* 56: 6960 and 7073. Brandegee made a very colorful argument against vocational education, saying that "after a man has had a common-school education in this country and can read . . . , if he can not learn how to dig the soil or run a lathe in a mill without some vocational educational artist visiting his home every day or two to see how he is getting along . . . well, then, we have [a] . . . paternal government, not the kind of government the fathers founded." He ended his rebuke claiming that the FBVE "is a molly-coddle kind of institution that will denature the whole American public if we keep up that sort of foolishness."

69. Senate Committee on Education and Labor, *Rehabilitation and Vocational Reeducation of Crippled Soldiers and Sailors*, S. Doc. 173, 61–68 and 27.
70. *Hearings on S. 4284 and H.R. 11367*, 72.
71. *Congressional Record* 56: 6955.
72. For more on Gallinger's role in the antivivisection campaign at the turn of the century, see Susan Lederer, *Subjected to Science: Human Experimentation in America before the Second World War* (Baltimore: Johns Hopkins University Press, 1995), and Susan Lederer, "Hideyo Noguchi's Leutin Experiment and Antivivisectionists," *Isis* 76, no. 1 (March 1985): 31–48. See also, Steven J. Smith et al., "Use of Animals in Biomedical Research: Historical Role of the American Medical Association and the American Physician," *Archives of Internal Medicine* 148 (August 1988): 1849–53.
73. *Congressional Record* 56: 7076.
74. Ibid., 7075.
75. *Hearings on S. 4284 and H.R. 11367*, 70.
76. Edward Stevens, "The Need of Better Hospital Equipment for the Medical Man," *Modern Hospital* 3 (December 1914): 367.
77. *Congressional Record* 56: 7075.
78. Ibid., 7073, 7076.
79. Grantham, *Hoke Smith*, 333.
80. As quoted in Berkowitz, "American Disability System," 43. According to Grantham the bill brought in a "modest appropriation of three million dollars." See Grantham, *Hoke Smith*, 333.
81. Daniel T. Rodgers and David B. Tyack "Work, Youth, and Schooling: Mapping Critical Research Areas," in Harvey Kantor and David B. Tyack, eds., *Work, Youth, and Schooling: Historical Perspectives on Vocationalism in American Education* (Stanford: Stanford University Press, 1982), 292.
82. Frank Billings, "The Relation of the Physician to Compulsory Sickness and Invalidity Insurance," *Journal of the American Medical Association* 46 (1906): 1471.
83. See Hoffman, *The Wages of Sickness*, and Jennifer Klein, *For All These Rights: Business, Labor, and the Shaping of America's Public-Private Welfare State* (Princeton, NJ: Princeton University Press, 2003).
84. *Hearings on S. 4284 and H.R. 11367*, 98.
85. Rosemary Stevens, *In Sickness and in Wealth*, 126.
86. The AMA House of Delegates made a resolution in the late 1920s to denounce the veteran health care system as unsound, "communistic" medicine. See Rosemary Stevens, *In Sickness and in Wealth*, 129.
87. For more on the White Committee, see Rosemary Stevens, "Can the Government Govern? Lessons from the Formation of the Veterans Administration," *Journal of Health Politics, Policy, and Law* 16.2 (Summer 1991): 281–305. See also Vanessa Northington Gamble, *Making a Place for Ourselves: The Black Hospital Movement, 1920–1945* (New York: Oxford University Press, 1995), esp. chap. 3, where Gamble discusses how the committee

endorsed the creation of the Tuskegee Veterans Hospital as a place where black ex-servicemen could receive health care segregated from white veterans.

88. Hickel, "Entitling Citizens," 126.

89. For US Treasury reports in 1920, see Edward D. Berkowitz and Kim Mc-Quaid, *Creating the Welfare State: The Political Economy of 20th-Century Reform*, rev. ed. (Lawrence: University Press of Kansas, 1992), 75. It is estimated that Congress faced annual expenditures of at least $10 billion dollars during this time period. One way that they attempted to manage the inflation in congressional spending was to institute the federal tax bill, placing an income tax on every citizen. See Watson, "A Testing Time."

EPILOGUE

1. David Polly, et al. "Advanced Medical Care for Soldiers Injured in Iraq and Afghanistan," *Minnesota Medicine* 87 no. 11 (Nov. 2004): 42–44.

2. Michael Weisskopf, "The Meaning of Walter Reed," *Time*, March 9, 2007, http://www.time.com/time/magazine/article/0,9171,1597533,00.html.

3. Prepared remarks of Secretary Rumsfeld for the BRAC Commission Hearing, May 16, 2005. http://www.defense.gov/Transcripts/Transcript .aspx?TranscriptID=3267.

4. Gordon Lubold, "How Decay Overtook Walter Reed," *Christian Science Monitor*, March 7, 2007, http://www.csmonitor.com/2007/0307/p01s01-usmi.html.

5. David Serlin also makes the argument that remasculinization trumps domestic and vocational rehabilitation in "Crippling Masculinity: Queerness and Disability in U.S. Military Culture, 1800–1945," *GLQ: A Journal of Lesbian and Gay Studies* 9 (2003): 168.

6. Mary Kaldor, *The Baroque Arsenal* (New York: Hill and Wang, 1981), 1.

7. Dana Priest and Anne Hull, "Soldiers Face Neglect, Frustration at Army's Top Medical Facility," *Washington Post*, February 18, 2007, A01.

8. For a sampling of historical work that addressed the Vietnam War, see Eric T. Dean, *Shook over Hell: Post-traumatic Stress, Vietnam, and the Civil War* (Cambridge, MA: Harvard University Press, 1997); Andrew J. Huebner, *The Warrior Image: Soldiers in American Culture from the Second World War to the Vietnam Era* (Chapel Hill: University of North Carolina Press, 2008); Susan Jeffords, *The Remasculinization of America: Gender and the Vietnam War* (Bloomington: Indiana University Press, 1989); and Paul Starr, *The Discarded Army: Veterans after Vietnam, The Nader Report on Vietnam Veterans and the Veterans Administration* (New York: Charterhouse, 1973).

9. George Vecsey, "A Veteran Unbowed by His War Injuries," *New York Times*, April 23, 2008, D1.

10. Anne Hull and Tamara Jones, "The War after the War: Soldiers' Battle Shifts

From Desert Sands to Hospital Linoleum," *Washington Post,* July 20, 2003, A01.

11. Ibid.

12. As quoted in Anne Hull and Tamara Jones, "The War after the War."

13. Available from http://georgewbush-whitehouse.archives.gov/news/releases /2003/12/20031218-1.html. See also Donna Miles, "DoD Allowing More Wounded Troops to Remain on Duty," American Forces Press Service, Oct. 14, 2004, http://www.globalsecurity.org/military/library/news/2004/10/ mil-041013-afps05.htm.

14. I first learned of the use of "tactical athletes" while I was engaged in a conversation with Joseph Miller, prosthetist, Military Advanced Training Center, who attended a presentation that I delivered at Walter Reed, summer 2008. Since then, I have noticed the usage of "tactical athletes" by many military officials involved in rehabilitation. See, for instance, Donna Miles, "New Center Offers Renewed Hope for Military Amputees," *American Forces Press Service,* San Antonio, Texas, Feb. 4, 2005, http://www.defense.gov/ news/newsarticle.aspx?id=25976.

15. See "New Military Center to Take Technology to the Next Level," http:// www.oandp.com/articles/2005-02_03.asp.

16. For war casualty figures, see http://www.globalsecurity.org/military/ops/ iraq_casualties.htm. For Walter Reed numbers, see Associated Press, "Amputees to Get New Rehab Center," November, 20, 2004, http://www .military.com/NewsContent/0,13319,FL_rehab_112004,00.html.

17. For reference to bionic warrior, see Sheri Waldrop and Michele Wojcie-chowski, "The 'Bionic' Warrior: Advances in Prosthetics, Technology, and Rehabilitation," *PT—Magazine of Physical Therapy* 15 (2007): 60–66, Earlier this was *P.T. Review.* http://www.apta.org/AM/Template.cfm?Section= Home&TEMPLATE=/CM/HTMLDisplay.cfm&CONTENTID=39050.

18. For Scoville's numbers, see Associated Press, "Amputees to Get New Rehab Center," November, 20, 2004, http://www.military.com/NewsContent/ 0,13319,FL_rehab_112004,00.html. Stansbury and colleagues, have concluded, however, that amputation rates for the wars in Iraq and Afghanistan are similar to those of previous conflicts; see Stansbury, et al., "Amputations in U.S. Military Personnel in the Current Conflicts in Afghanistan and Iraq," *Journal of Orthopedic Trauma* 22.1 (January 2008): 43–46.

19. US Army Corps of Engineers, "FY04 Military Construction Project Data," September 2004. Elihu Hirsch, project manager for the US Army Corps of Engineers, kindly sent me this document. Elihu Hirsch email message to author, March 3, 2009.

20. US Army Corps of Engineers, "FY04 Military Construction Project Data," September 2004. In author's possession.

21. David Serlin, "Engineering Masculinity: Veterans and Prosthetics after

World War Two," in Katherine Ott, David Serlin, and Stephen Mihm, eds., *Artificial Parts, Practical Lives: Modern Histories of Prosthetics* (New York: New York University Press, 2002), 54.

22. Michael Belfiore, *The Department of Mad Scientists: How DARPA Is Remaking Our World, from the Internet to Artificial Limbs* (Washington, DC: Smithsonian Books, 2009).

23. David Serlin "Engineering Masculinity," 55.

24. Belfiore, *The Department of Mad Scientists*, xx.

25. Statement of Dr. Brett Giroir, Deputy Director, Defense Sciences Office, Defense Advanced Research Projects Agency, statement posted July 22, 2004. http://veterans.house.gov/hearings/schedule108/ju104/7-22-04/bgiroir .html.

26. According to a 1927 report published by the Surgeon General's Office, "less than 20 percent of persons with amputations of the upper arm" from the First World War wore prosthetic limbs. See *MDWW*, vol. 11, pt. 1, 737.

27. Take, for instance, the story of Dawn Halfaker in Donna St. George, "Limbs Lost to Enemy Fire, Women Forge a New Reality," *Washington Post*, April 18, 2006, A01.

28. Michael Weisskopf, *Blood Brothers: Among the Soldiers of Ward 57* (New York: Henry Holt and Company, 2006), 125–26.

29. Giroir, Defense Advanced Research Projects Agency, http://veterans.house .gov/hearings/schedule108/ju104/7-22-04/bgiroir.html.

30. Not that Congress needed much coaxing. Many congressmen had been to Ward 57. Debates on the House floor invoked World War I British poet and soldier-amputee, Wilfred Owen. See Representative Adam B. Schiff of California, "The Debt We Owe to Our Wounded," on October 9, 2004, 108th Cong., 2nd sess., *Congressional Record* 150, E1961–E1962.

31. Steve Vogel, "Military Hospitals Meet New Realities," *Washington Post*, Feb. 11, 2008, http://www.washingtonpost.com/wp-dyn/content/article/2008/ 02/10/AR2008021002214.html?sid=ST2008021100035.

32. As quoted in Tom Ramstack "Reed to Build Amputee Center," *Washington Times*, Wednesday, Aug. 3, 2006, C10.

33. Craig Coleman, "Military Advanced Training Center Opens at WRAMC," *Army News Service* Sept. 13, 2007, http://www.military.com/features/ 0,15240,149156,00.html.

34. Amanda Kolson Hurley, "Walk This Way: Walter Reed's New Gait Lab Helps Soldiers Regain Mobility Post-Combat," *Architect Magazine*, April 1, 2008, http://www.architectmagazine.com/concrete-construction/walk-this-way .aspx.

35. Steve Vogel, "For War's Wounded, Space to Heal," *Washington Post*, Sept. 13, 2007, B01.

36. Jim Garamone, "Congress Sends $416.2 Billion Budget to President," *American Forces Press Service*, July 26, 2004, http://www.defense.gov/news/ ju12004/.

37. Elihu Hirsch email message to author, March 3, 2009.

38. Soldiers fighting for the US during the First World War were, for the most part, drafted through the 1917 Selective Service Act. At the time, at least one-third of draft-age men were found to be "unfit," and even those who made it into uniform were seen as marginally fit. World War I rehabilitation officials often felt they were working with the dregs of society, gangly, disorderly men who needed to be physically and mentally disciplined. For more, see Jennifer Keene, *Doughboys, the Great War, and the Remaking of America* (Baltimore: Johns Hopkins University Press, 2001). Today, by contrast, soldiers deployed to Iraq and Afghanistan are career soldiers who are understood to be in peak physical shape. Rehabilitators refer to their soldier-patients as "professional military athletes," who enjoy high-impact extreme sports such as snowboarding and motor cross.

39. See the example of Roberto Reyes in Dana Priest, "Call God for Help," Washington Post online. http://www.washingtonpost.com/wp-srv/photo/galleries/070216/walterreed/index.html?tab=0&gal=day1.

40. Anne Hull, "Wounded or Disabled But Still on Active Duty," *Washington Post*, December 1, 2004, A23.

41. Matt Mireles, "Garth Stewart: The Untroubled Soldier," http://www.mattmireles.com/Garth%20Stewart.html.

42. As of April 2006 there were 11 female amputees compared to 350 male amputees. See Donna St. George "Limbs Lost to Enemy Fire, Women Forge a New Reality," *Washington Post*, April 18, 2006, A01.

43. Ibid.

44. Matthew Clifton, "Amputee Achieves Goal: Returns to Iraq," *American Forces Press Service*, April 12, 2005, http://www.defense.gov/news/newsarticle.aspx?id=31438.

45. "Murtha Dedicates Walter Reed Military Advanced Training Center," *US Fed News Service*, Sept. 13, 2007, http://www.house.gov/list/press/pa12_murtha/wramputee.html.

46. Angie Cannon, "The Front Lines of Healing: When Injured Soldiers Come Home," *U.S. News and World Report*, July 28, 2003, 74, 76.

47. Weisskopf, *Blood Brothers*, 221.

Bibliography

Selected Archival Sources

American Physical Therapy Association Archives, Alexandria, VA
Carlisle Barracks, Archives and Manuscripts, Carlisle, Pennsylvania
    Fort Sheridan Museum Collection, 1865–1987
    H. Winnett Orr Papers
    W. Don Jones Papers
    World War I Veterans Survey
The College of Physicians of Philadelphia, Philadelphia, PA
    Philadelphia Orthopedic Hospital and Infirmary for Nervous
        Disease, 1867–1942
    Anderson Hospital Collection
    George W. Outerbridge Papers
    S. B. Sturgis World War I Records
National Archives and Records Administration, College Park, MD
    Record Group 112, Records of the Office of the Surgeon
        General (Army)
    Record Group 159, Records of the Office of the Inspector
        General (Army)
    Record Group 200, Records of the American Red Cross,
        1917–1934, National Archives Gift Collection
National Archives and Records Administration, Washington, DC
    Record Group 15, Records of the Department of Veterans
        Affairs
National Library of Medicine, National Institutes of Health,
    Bethesda, MD
National Museum of Health and Medicine, Otis Historical Ar-
    chives, Washington, DC
    Angier and Hitchcock Collection
    Army Medical Museum Photographs
    Vogel Collection
    Walter Reed Army Medical Center History Collection

National World War I Museum, Kansas City, MO
New York Academy of Medicine, Historical Collections, New York, NY
   Douglas McMurtrie Collection

Periodicals

American Forces Press Services
American Journal of the Care of Cripples
American Journal of Orthopedic Surgery
Annals of the American Academy of Political and Social Science
Architect Magazine
Carry On: A Magazine on the Reconstruction of Disabled Soldiers and Sailors
Century Illustrated
Christian Science Monitor
Columbia Daily Spectator
The Come-Back
Harper's
Journal of the American Medical Association
The Los Angeles Times
The Military Surgeon
The Nation
New York Medical Journal
The New Republic
The New York Times
Outlook
The Pennsylvania Medical Journal
P.T. Review
Physiotherapy Review
Transactions of the American Orthopaedic Association
Transactions of the College of Physicians of Philadelphia
US News and World Report
The Washington Herald
The Washington Post
World's Work

Congressional Records

U.S. Congress. Congressional Record. 65th Cong., 2nd sess., May–June 1918.
   Vol. 56, pt. 7.

———. House. Committee on Education. *Vocational Rehabilitation of Disabled Soldiers and Sailors: Hearings on H.R. 11367.* 65th Cong., 2nd sess., April 30, May 1, 2, 1918.

———. Senate. Committee on Education and Labor. *Vocational Rehabilitation of Disabled Soldiers and Sailors: Hearings on S. 4284.* 65th Cong., 2nd sess., April 30, May 1, 2, 1918.

———. Senate. Committee on Education and Labor. *Vocational Rehabilitation of Disabled Soldiers and Sailors.* 65th Cong., 2nd sess., January 1918. S. doc. 166.

———. Senate. Committee on Education and Labor. *Rehabilitation and Vocational Reeducation of Crippled Soldiers and Sailors.* 65th Cong., 2nd sess., February 1918. S. Doc. 173.

Primary Sources

Addams, Jane. *My Friend, Julia Lathrop.* New York: Macmillan, 1935.

Albee, Fred H. "The Function of the Military Orthopedic Hospital." *New York Medical Journal* 106 (1917): 2–4.

———. *A Surgeon's Fight to Rebuild Men; an Autobiography.* New York: Dutton, 1943.

Army Surgeon General's Office. *Military Orthopaedic Surgery.* 2nd ed. Philadelphia: Lea & Febiger, 1918.

Ashburn, P. M. *A History of the Medical Department of the United States Army.* Boston and New York: Houghton Mifflin Company, 1929.

Baldwin, Bird T. *Occupational Therapy Applied to Restoration of Movement.* Washington, DC: Walter Reed General Hospital, 1919.

———. "The Function of Psychology in the Rehabilitation of Disabled Soldiers." *Psychological Bulletin* 16 (August 1919): 267–90.

Bartlett, Helen Conkling. "The Teaching of Massage to Pupils in Hospital Training-Schools." *American Journal of Nursing* (1901): 718–21.

Billings, Frank. "The Relation of the Physician to Compulsory Sickness and Invalidity Insurance." *Journal of the American Medical Association* 46 (1906): 1470–72.

Borden, William C. "The Walter Reed General Hospital of the United States Army." *Military Surgeon* 20 (1907): 20–35.

Brackett, Elliot. "Meeting of the Boston Orthopedic Club, Oct. 6, 1717." *American Journal of Orthopaedic Surgery* 15 (1917): 842–56.

Bradford, Edward, and Robert Lovett. *A Treatise on Orthopedic Surgery.* 1st ed. New York: William Wood & Co., 1890.

Brown, John Elliott, and Edward F. Stevens. "General Hospital for One Hundred Patients." In Charlotte A. Aikens, ed., *Hospital Management,* 108–47. Philadelphia: Saunders, 1911.

Carnes Artificial Limb Company. *The Carnes Arm Puts You on the Pay Roll.* Kansas City: Chas E. Brown Printing Co, ca. 1913.

Davis, Gwilym G. "President's Address." *American Journal of Orthopaedic Surgery* 12 (July 1914): 1–4.

———. "The Relation of Cripples to the Public." *Transactions of the College of Physicians of Philadelphia* 39 (1917): 476–84.

Decker, Ruby. "Recollections and Reminiscences from Former Reconstruction Aides." *Physical Therapy* 56 (January 1976): 25.

Defense Advanced Research Projects Agency. http://veterans.house.gove/hearings/schedule108/ju104/7-22-04/bgiroir.html.

Devine, Edward T. *Disabled Soldiers and Sailors Pensions and Training.* New York: Oxford University Press, 1919.

Dewey, John. *Democracy and Education: An Introduction to the Philosophy of Education.* New York: Macmillan Co., 1916.

Dock, L. L. "Is the Profession Becoming Overcrowded?" *American Journal of Nursing* 3 (April 1903): 513–15.

Dock, L. L., Anna R. Rutherford, and Helen Kelly. "Letters to the Editor." *American Journal of Nursing* 2 (1901): 62–66

Douglas, P. H. "The War Risk Insurance Act." *Journal of Political Economy* 26 (1918): 461–83.

Ewing, James. "Experiences in the Collection of Museum Material from Army Camp Hospitals." *Bulletin of the International Association of Medical Museums* 8 (December 1922): 27–34.

Faries, John Culbert. *The Economic Consequences of Physical Disability: A Case Study of Civilian Cripples in New York City.* New York City: Red Cross Institute for Crippled and Disabled Men, 1918.

Federal Board for Vocational Education. *Vocational Rehabilitation of Disabled Soldiers and Sailors: A Preliminary Study.* Washington, DC: GPO, January, 1918.

———. *Taking His Place in Industry: How the Disabled Soldier and Sailor Will Be Put in a Good Job.* New York: National Security League, 1919.

Franz, S. I. "Reeducation and Rehabilitation of Crippled, Maimed and Otherwise Disabled by War." *Journal of the American Medical Association* LXIX (7 July 1917): 63–64.

Gibson, Henry Richard. *Liberality, Not Parsimony: This Is Our True Pension Policy.* Washington, DC: GPO, 1900.

Glasson, William Henry. *Federal Military Pensions in the United States.* New York: Oxford University Press, 1918.

Goldthwait, Joel E. "Dislocation of the Patella," *Transactions of the American Orthopaedic Association* 8 (1896): 237–38.

———. "With the Report of Eleven Cases Slipping or Recurrent Dislocation of the Patella." *American Journal of Orthopaedic Surgery* 2 (1904): 293–308.

———. "The Place of Orthopedic Surgery in War" *American Journal of Orthopaedic Surgery* 15 (1917): 679–86.

———. "The Orthopedic Preparedness of the Nation" *American Journal of Ortho-paedic Surgery* 15 (1917): 219–20.

———. *The Division of Orthopaedic Surgery in the A.E.F.* Norwood, MA: Plimpton Press, 1941.

Harris, Garrard. *The Redemption of the Disabled: A Study of Programmes of Rehabili-tation for the Disabled of War and Industry.* New York: D. Appleton, 1919.

Hazenhyer, Ida May. "A History of the American Physiotherapy Association." *Physiotherapy Review* 26 (February 1946): 3–14, 66–74, 122–29, 174–84.

Hendrick, Burton J. "Pork-Barrel Pensions." *World's Work* 30 (October 1915): 713–20.

Heroes All. Produced and directed by the American Red Cross Bureau of Pictures. 9 min. John E. Allen Inc., ca.1918. Videocasette.

Hibbs, R. A. "An Operation for Progressive Spinal Deformities." *New York Medical Journal* 93 (1911): 1013–16.

History of Letterman General Hospital. San Francisco, CA: Listening Post, 1919.

"Immoral 'Massage' Establishment," *British Medical Journal* July 14, 1894: 88.

The Inter-Allied Conference on the After-Care of Disabled Men. London: His Majesty's Stationery Office, 1918.

Knight, James. *Orthopaedia or a Practical Treatise on the Aberrations of the Human Form.* New York: G. P. Putnam's Sons, 1874.

Lathrop, Julia C. "The Military and Naval Insurance Act." *Nation* 106 (February 7, 1918): 157–58.

Law, DeWitt. *Soldiers of the D.A.V: A History of the Disabled War Veterans and the American Pension System.* DeWitt Law, 1929.

Lindsay, S. M. C. "Soldiers' and Sailors' Insurance." *Proceedings of the American Philosophical Society* 57 (1918): 632–48.

———. "Purpose and Scope of War Risk Insurance." *Annals of the American Acad-emy of Political and Social Science* 79 (1918): 52–68.

Little, E. Muirhead. "Notes on Artificial Limbs for Sailors and Soldiers." *American Journal of Orthopedic Surgery* 15 (1917): 596–602.

Love, T. B. "The Social Significance of War Risk Insurance." *Annals of the Ameri-can Academy of Political and Social Science* 79 (1918): 46–51.

Lovett, Robert W. "The Treatment of Pott's Disease" *Transactions of the American Orthopaedic Association* 8 (1896): 128–52.

Marks, A. A. *Manual of Artificial Limbs: An Exhaustive Exposition of Prothesis.* New York: A. A. Marks, Inc. 1926.

Maverick, Maury. *A Maverick American.* New York: Convici Friede, 1937.

Mayer, Leo. "The Organization and Aims of the Orthopedic Reconstruction Hos-pital." *American Journal of the Care for Cripples* 5 (September 1917): 78–85.

———. "The Military Orthopedic Reconstruction Hospital." *Journal of the Ameri-can Medical Association* 69, no. 18 (November 3, 1917): 1522–24

McDill, John R. *Lessons from the Enemy: How Germany Cares for Her War Disabled.* Medical War Manual, no. 5. Philadelphia and New York: Lea & Febiger, 1918.

McKenzie, R. Tait. *Reclaiming the Maimed: A Handbook of Physical Therapy*. New York: Macmillan, 1918.

McMillan, Mary. *Massage and Therapeutic Exercise*. Philadelphia: W. B. Saunders Company, 1921.

McMurry, Donald L. "The Political Significance of the Pension Question, 1885–1897." *Mississippi Valley Historical Review* 9 (June 1922): 19–36.

McMurtrie, Douglas. "The Care of Crippled Children in the United States." *American Journal of Orthopedic Surgery* 9 (1912): 527–56.

———. *An American Program for the Rehabilitation of Disabled Soldiers*. New York: Red Cross Institute for the Crippled and Disabled Men, 1918.

———. "War-Cripple Axioms and Fundamentals." *Columbia University War Pamphlet* 17 (November, 1917): 65.

———. *The Meaning of the Term "Crippled."* New York: William Wood & Co., 1918.

———. *The Organization, Work, and Method of the Red Cross Institute for Crippled and Disabled Men*. New York: Red Cross Institute for Crippled and Disabled Men, 1918.

———. *The Disabled Soldier*. New York: Macmillan, 1919.

———. *A Graphic Exhibit on Rehabilitation of the Crippled and the Blinded*. New York: Red Cross Institute for the Crippled and Disabled Men, 1919.

———. "The Influence of Pension or Compensation Administration on the Rehabilitation of Disabled Soldiers." *American Medicine* 14, no. 6 (1919): 355–65.

Meriam, Lewis, and Karl T. Schlotterbeck. *The Cost and Financing of Social Security*. Washington: Brookings Institution, 1950.

Miller, Alice Duer. "How Can a Woman Best Help?" *Carry On: A Magazine on the Reconstruction of Disabled Soldiers and Sailors* 1 (June 1918): 17–18.

Mireles, Matt. "Garth Stewart: The Untroubled Soldier." Matt Mireles Blog. http://www.mattmireles.com/Garth%20Stewart.html (accessed May 16, 2010)

Mock, Harry E. "The Disabled in Industry." *Annals of the American Academy of Political and Social Science* 80 (November 1918): 29–34.

Mudd, L. C. "The Letterman General Hospital." *New York Medical Journal* 109 (1919): 242–44.

Munson, Edward Lyman. *The Soldier's Foot and the Military Shoe: A Handbook for Officers and Noncommissioned Officers of the Line*. Menasha, Wisconsin: George Banta Publishing Co., 1912.

National Research Council. *Human Limbs and Their Substitutes, Presenting Results of Engineering and Medical Studies of the Human Extremities and Application of the Data to the Design and Fitting of Artificial Limbs and to the Care and Training of Amputees*. New York: McGraw-Hill, 1954.

"News Notes," *American Physical Education Review* 20 (January 1915): 42–43.

Nissen, Hartwig. "The Swedish Movement and Massage Treatment." *Journal of the American Medical Association* 10 (April 7, 1888): 423–24.

Office of the Surgeon General. *Defects Found in Drafted Men*. Washington, DC: GPO, 1920.

———. *The Medical Department of the United States Army in the World War.* Vols. 1–15. Washington, DC: GPO, 1921–29.

———. *Army Medical Specialist Corps*. Washington, DC: GPO, 1968.

Oliver, John William. "History of the Civil War Military Pensions, 1861–1885." *Bulletin of the University of Wisconsin,* no. 844, History Series, no. 1 (1917): 1–120.

Orr, H. Winnett. "The Industrial Education of the Crippled and Deformed." *American Journal of Orthopedic Surgery* 10 (1912): 195–200.

Osgood, Robert Bayley. "A Survey of the Orthopaedic Services in the U.S. Army Hospitals, General Base, and Debarkation." *American Journal of Orthopedic Surgery* 17 (1919): 359–82.

———. *The Evolution of Orthopaedic Surgery*. St. Louis: C. V. Mosby Company, 1925.

Owen, W. O. "The Army Medical Museum." *New York Medical Journal* 107 (June 1918): 1034–36.

Pinel, Philippe. *A Treatise on Insanity*. Sheffield: Todd, 1806.

Polly, D. W., Jr., T. R. Kuklo, W. C. Doukas, and C. Scoville. "Advanced Medical Care for Soldiers Injured in Iraq and Afghanistan." *Minnesota Medicine* 87, no. 11 (Nov. 2004): 42–44.

President's Commission on Veterans' Pensions. *The Historical Development of Veterans' Benefits in the United States: A Report on Veterans' Benefits in the United States*. Washington, DC: GPO, 1956.

Reeves, Edith. *Care and Education of Crippled Children in the United States*. New York: Russell Sage Foundation, 1914.

"Rest When You're Dead: The Story of Garth Stewart." Frontrunner Magazine. http://www.frontrunnermagazine.com/2009/10/08/rest-when-youre-dead-the-story-of-garth-stewart/ (accessed May 16, 2010).

Rubinow, I. M. *Social Insurance*. 1913; New York: Arno, 1969.

Sanderson, Marguerite. "Some Aspects of the Massage Problem," *American Journal of Orthopedic Surgery* 16 (1918): 437.

Silver, David. "The Etiological Importance of Abnormal Foot Posture in Affections of the Knee." *American Journal of the Medical Sciences* 136 (1908): 726–34.

———. "The Care of the Amputated in War." *Pennsylvania Medical Journal* 22 (January 1919): 212–15.

———. "Lessons from the War in the Care of Industrial Injuries of the Extremities." *West Virginia Medical Journal* 14 (September 1919): 86–90.

Sloane, William M. "Pensions and Socialism." *Century Illustrated Magazine* 42 (June 1891): 179–89.

Southard, W. F. "Opening of the New Military Hospital at the Presidio." *Pacific Medical Journal* 42 (July 9, 1899): 454–64.

Stansbury, Lynn G., Steven J. Lalliss, Joanna G. Branstetter, Mark R. Bagg, and John B. Holcomb. "Amputations in U.S. Military Personnel in the Current Conflicts in Afghanistan and Iraq." *Journal of Orthopedic Trauma* 22.1 (January 2008): 43–46.

Starr, Clarence L. "The Role of Orthopedic Surgery in Modern Warfare." *American Journal of Orthopedic Surgery* 16 (1918): 415–26.

Stevens, Edward. "The Need of Better Hospital Equipment for the Medical Man." *Modern Hospital* 3 (December 1914): 367–71.

———."The Physiotherapy Department for the American Hospital." *Architectural Record* 60 (July 1926): 18–24.

Taylor, Charles Fayette. "The Spinal Assistant: Autobiographical Reminiscences." *Transactions of the American Orthopaedic Association* 12 (1899): 15–27.

"The Scandals of Massage: Report on the Special Commissioners of the British Medical Journal." *British Medical Journal,* Nov. 24, 1894, 1199–1200.

Thomas, Leah C., and Joel E. Goldthwait. *Body Mechanics and Health.* Boston: Houghton Mifflin Co., 1922.

Vogel, Emma. "Physical Therapists before WWII." In Robert S. Anderson, Harriet S. Lee, and Myra L. McDaniel, eds., *Army Medical Specialist Corps.* Washington, DC: Office of the Surgeon General, 1968.

Warner, Arthur. "The Betrayal of Our War Victims." *Nation* 118, no. 3061 (March 5, 1924): 249–50.

Waterbury, C. Harold. "Modern Forms of Insurance Protection." *Nation* 106, no. 2745 (February 7, 1918): 164–65.

Weber, Gustavus Adolphus. *The Employees' Compensation Commission: Its History, Activities, and Organizations.* New York and London: D. Appleton and Company, 1922.

Weisskopf, Michael. "The Meaning of Walter Reed." *Time,* March 9, 2007.

Whitman, Royal. "A Definition of the Scope of Orthopaedic Surgery, as Indicated by its Origins, its Development, and by the Work of the American Orthopaedic Association." *Transactions of the American Orthopaedic Association* 8 (1896): 1–9.

———. "Critical Estimation of the Personal Influence of Four Pioneers on the Development of Orthopaedic Surgery in New York." *Journal of Bone Joint Surgery of America* 16 (1934): 331–42.

Willard, DeForest. "Children's Orthopedic Ward, Agnew Memorial Pavilion," *Transactions of the American Orthopaedic Association* 11 (1898): 456–61.

———. "What Shall We Do with Our Cripples?" *Journal of the American Medical Association* 52 (April 3, 1909): 1134–36.

Wolfe, S. H. "Eight Months of War Risk Insurance Work." *Annals of the American Academy of Political and Social Science* 79 (1918): 68–79.

Wood, Casey. "Carry On." *American Journal of Orthopedic Surgery* 16, no. 8 (August 1918): 539.

Woodward, Lemuel F. "Orthopedic Appliance Shop." *Transactions of the American Orthopaedic Association* 11 (1898): 76–82.

Secondary Sources

Abel, Emily K. "Valuing Care: Turn-of-the-Century Conflicts between Charity Workers and Women Clients." *Journal of Women's History* 10, no. 3 (Autumn, 1998): 32–52.

Adams, Annmarie. "Borrowed Buildings: Canada's Temporary Hospitals during World War I." *Canadian Bulletin of the History of Medicine* 16 (1999): 25–48.

———. *Medicine by Design: The Architect and the Modern Hospital, 1893–1943.* Minneapolis: University of Minnesota Press, 2008.

Aldrich, Mark. *Safety First: Technology, Labor, and Business in the Building of American Work Safety, 1870–1939.* Baltimore: Johns Hopkins University Press, 1997.

———. "Train Wrecks to Typhoid Fever: The Development of Railroad Medicine Organizations, 1850 to World War I." *Bulletin of the History of Medicine* 75, no. 2 (2001): 254–89.

Amirault, Chris. "Posing the Subject of Early Medical Photography." *Discourse* 16, no. 2 (Winter 1993–94): 51–76.

Anderson, Julie, Francis Neary, and John V. Pickstone. *Surgeons, Manufacturers and Patients: A Transatlantic History of Total Hip Replacement.* New York: Palgrave Macmillan, 2007.

Anderson, Julie, and Neil Pemberton. "Walking Alone: Aiding the War and the Civilian Blind in the Inter-war Period." *European Review of History* 14, no. 4 (December 2007): 459–79.

Apple, Rima. "Image or Reality? Photographs in the History of Nursing." In Anne Hudson Jones, ed., *Images of Nurses: Perspectives from History, Art, and Literature.* Philadelphia: University of Pennsylvania Press, 1988.

———. *Vitamania: Vitamins in American Culture.* New Brunswick, NJ: Rutgers University Press, 1996.

———. *Perfect Motherhood: Science and Childrearing in America.* New Brunswick, NJ: Rutgers University Press, 2006.

Apple, Rima, and Janet Golden. *Mothers and Motherhood: Readings in American History.* Columbus: Ohio State University Press, 1997.

Atkinson, Paul. "The Feminist Physique: Physical Education and the Medicalization of Women's Education." In J. A. Mangan and Roberta J. Park, eds., *From 'Fair Sex' to Feminism: Sport and the Socialization of Women in the Industrial and Post-Industrial Eras,* 38–57. London: Frank Cass & Co., 1987.

Baker, Robert, ed. *The American Medical Ethics Revolution: How the AMA's Code of Ethics Has Transformed Physicians' Relationships to Patients, Professionals, and Society.* Baltimore: Johns Hopkins University Press, 1999.

Barclay, Jean. *In Good Hands: The History of the Chartered Society of Physiotherapy, 1894–1994.* Oxford: Butterworth-Heinemann, 1994.

Barnard, H. *The Forging of an American Jew: The Life and Times of Judge Julian W. Mack.* New York: Herzl Press, 1974.

Barnes, David. *The Great Stink of Paris and the Nineteenth-Century Struggle against Filth and Germs.* Baltimore: Johns Hopkins University Press, 2006.

Bates, Barbara. *Bargaining for Life: A Social History of Tuberculosis, 1876–1938.* Philadelphia: University of Pennsylvania Press, 1992.

Bean, Bennett William. *Walter Reed: A Biography.* Charlottesville: University Press of Virginia, 1982.

Beardsley, Edward H. *A History of Neglect: Health Care for Blacks and Mill Workers in the Twentieth-Century South.* Knoxville: University of Tennessee Press, 1987.

Bederman, Gail. *Manliness and Civilization: A Cultural History of Gender and Race in the United States, 1880–1917.* Chicago: University of Chicago Press, 1995.

Belfiore, Michael. *The Department of Mad Scientists: How DARPA Is Remaking Our World, from the Internet to Artificial Limbs.* Washington, DC: Smithsonian Books, 2009.

Berkowitz, Edward D. *Disabled Policy: America's Programs for the Handicapped.* Cambridge: Cambridge University Press, 1987.

———, ed. *Disability Policies and Government Programs.* New York: Praeger, 1979.

Berkowitz, Edward D., and Kim McQuaid. *Creating the Welfare State: The Political Economy of 20th-Century Reform.* Rev. ed. Lawrence: University Press of Kansas, 1992.

Blanck, Peter David, and Michael Millender. "Before Disability Civil Rights: Civil War Pensions and the Politics of Disability in America." *Alabama Law Review* 52, no. 1 (2000): 1–50.

Bland, Lucy. *Banishing the Beast: Feminism, Sex and Morality.* New York: New Press, 1995.

Blight, David W. *Race and Reunion: The Civil War in American Memory.* Cambridge, MA: Belknap Press of Harvard University Press, 2001.

———. *Beyond the Battlefield: Race, Memory and the American Civil War.* Amherst: University of Massachusetts Press, 2002.

Bourke, Joanna. *Dismembering the Male: Men's Bodies, Britain and the Great War.* London: Reaktion Books, 1996.

Braeman, John. "World War One and the Crisis of American Liberty." *American Quarterly* 15 (Spring 1964): 104–12.

Brandt, Allan M. *No Magic Bullet: A Social History of Venereal Disease in the United States.* New York: Oxford University Press, 1987.

———. "The Cigarette, Risk, and American Culture." *Daedalus* 119 (Fall 1990): 155–76.

Brandt, Allan M., and Martha Gardner. "The Golden Age of Medicine?" In Roger Cooter and John Pickstone, eds., *Companion to Medicine in the Twentieth Century.* London: Routledge, 2000.

Bristow, Nancy. *Making Men Moral: Social Engineering During the Great War.* New York: New York University Press, 1996.

Bogdan, Robert. *Freak Show: Presenting Human Oddities for Amusement and Profit.* Chicago: University of Chicago, 1988.

Brieger, Gert H. "A Portrait of Surgery: Surgery in America, 1875–1889." *Surgical Clinics of North America* 67 (December 1987): 1181–1215.

Brody, David, *Workers in Industrial America: Essays on the Twentieth-Century Struggle.* 2nd ed. New York: Oxford University Press, 1993.

Brown, Bill. "Science Fiction, the World's Fair, and the Prosthetics of Empire, 1910–1915." In Amy Kaplan and Donald E. Pease, eds., *Cultures of United States Imperialism.* Durham: Duke University Press, 1993.

Brown, Richard E. *Rockefeller Medicine Men: Medicine and Capitalism in America.* Berkeley: University of California Press, 1979.

Brown, Thornton. *The American Orthopaedic Association: A Centennial History.* Park Ridge, IL: American Orthopaedic Association, 1987.

Bryder, Linda. *Below the Magic Mountain: A Social History of Tuberculosis in Twentieth-Century Britain.* Oxford: Clarendon Press, 1988.

Burch, Susan, and Hannah Joyer. *Unspeakable: The Story of Junius Wilson.* Chapel Hill: University of North Carolina Press, 2007.

Burch, Susan, and Ian Sutherland. "Who's Not yet Here? American Disability History." *Radical History Review* 94 (2006): 127–47.

Burnham, John C. "American Medicine's Golden Age: What Happened to It?" *Science* 215 (19 March 1982): 1474–79.

Byerly, Carol R. *Fever of War: The Influenza Epidemic in the U.S. Army during World War I.* New York: New York University Press, 2005.

———. *Good Tuberculosis Men: The Army Medical Department's Struggle with Tuberculosis.* Washington, DC: Borden Institute, forthcoming.

Byrom, Bradley Allen. "A Pupil and a Patient: Hospital-Schools in Progressive America." In Paul K. Longmore and Lauri Umansky, eds., *The New Disability History: American Perspectives.* New York: New York University Press, 2001.

———. "A Vision of Self Support: Disability and the Rehabilitation Movement in Progressive America." PhD diss., University of Iowa, 2004.

Calhoun, C. W. "Reimagining the 'Lost Men' of the Gilded Age: Perspectives on the Late Nineteenth Century Presidents." *Journal of the Gilded Age and Progressive Era* 1, no. 3 (July 2002): 225–57.

Canguilhem, Georges. "Machine and Organism." In Jonathan Crary and Sanford Kwinter, eds., *Incorporations.* New York: Zone, 1992.

Carden-Coyne, Ana. "Ungrateful Bodies: Rehabilitation, Resistance, and Disabled American Veterans of the First World War." *European Review of History* 14, no. 4 (2007): 543–65.

———. "Painful Bodies and Brutal Women: Remedial Massage, Gender Relations and Cultural Agency in Military Hospitals, 1914–18." *Journal of War and Culture Studies* 1, no. 2 (2008) 139–58.

———. *Reconstructing the Body: Classicism, Modernism, and the First World War.* Oxford: Oxford University Press, 2009.

Carmichael, Jane. *First World War Photographers.* London: Routledge, 1989.

Carter, M. "The Early Days of Orthopedic Nursing in the United Kingdom: Agnes Hunt and Baschurch." *Orthopaedic Nursing* 19 (2000): 15–18.

Carter, Susan B., ed. *Historical Statistics of the United States,* vol. 1. Cambridge: Cambridge University Press, 2006.

Charlton, James I. *Nothing about Us without Us: Disability Oppression and Empowerment*. Berkeley: University of California Press, 2000.

Clements, Kendrick A. *The Presidency of Woodrow Wilson*. American Presidency series;. Lawrence: University Press of Kansas, 1992.

Cohen, Deborah. *The War Come Home: Disabled Veterans in Britain and Germany, 1914–1939*. Berkeley: University of California Press, 2001.

Connolly, Cynthia. *Saving Sickly Children: The Tuberculosis Preventorium in American Life, 1909–1970*. New Brunswick, NJ: Rutgers University Press, 2008.

Connor, J. T. H., and Michael G. Rhode. "Shooting Soldiers: Civil War Medical Images, Memory, and Identity in America." *Invisible Culture: An Electronic Journal for Visual Culture Issue* 5 (2003) http://www.rochester.edu/in_visible.

Conrad, Peter. *Medicalization of Society: On the Transformation of Human Conditions into Treatable Disorders*. Baltimore: Johns Hopkins University Press, 2007.

Conrad, Peter, and Joseph Schneider. *Deviance and Medicalization: From Badness to Sickness*. 2nd ed. Philadelphia: Temple University Press, 1992.

Cooper, John Milton, Jr. *Walter Hines Page: The Southerner as American, 1855–1918*. The Fred W. Morrison series in Southern studies;. Chapel Hill: University of North Carolina Press, 1977.

Cooper, David Y. "The Evolution of Orthopaedic Surgeons from Bone and Joint Surgery at the University of Pennsylvania." *Clinical Orthopaedics and Related Research* 374 (May 2000): 17–35.

Cooter, Roger. *Surgery and Society in Peace and War: Orthopaedics and the Organization of Modern Medicine, 1880–1948*. London: Macmillan Press, 1993.

———. "The Disabled Body." In Roger Cooter and John Pickstone, eds., *Companion to Medicine in the Twentieth Century*. London: Routledge, 2000.

———, ed. *In the Name of the Child: Health and Welfare, 1880–1940*. London: Routledge, 1992.

Cooter, Roger, and Bill Luckin, eds. *Accidents in History: Injuries, Fatalities and Social Relations*. Amsterdam: Rodopi, 1997.

Cooter, Roger, Mark Harrison, and Steve Sturdy, eds. *War, Medicine and Modernity*. Stroud: Sutton Publishing, 1998.

Cott, Nancy. *The Grounding of Modern Feminism*. New Haven: Yale University Press, 1987.

Creer, Ralph P. "Medical Illustration in the United States Army: Historical and Present Conditions." *Journal of Laboratory and Clinical Medicine* 28 (February 1943): 651–61.

Crnic, Meghan and Cynthia Connolly. "'They Can't Help Getting Well Here': Seaside Hospital for Children in the United States, 1872–1917." *Journal of the History of Childhood and Youth* 2 (2009): 220–33.

Crosby, Alfred. *America's Forgotten Pandemic: The Influenza of 1918*. New York: Cambridge University Press, 1989.

Curtin, Roland G. "Memoir of DeForest Willard." *Transactions of the College of Physicians Philadelphia* 33 (1911): 74–84.

Daily, Maceo Crenshaw Jr. "Neither 'Uncle Tom' nor 'Accommodationist': Booker T. Washington, Emmet Jay Scott, and Constitutionalism." *Atlanta History* 38, no. 4 (1995): 20–33.

D'Antonio, Patricia. *Founding Friends: Families, Staff, and Patients at the Friends Asylum in Early Nineteenth-Century Philadelphia*. Bethlehem, PA: Lehigh University Press, 2006.

———. *American Nursing: A History of Knowledge, Authority, and the Meaning of Work*. Baltimore: Johns Hopkins University Press, 2010.

Davies, Wallace Evan. *Patriotism on Parade: The Story of Veteran's and Hereditary Organizations in America, 1783–1900*. Cambridge, MA: Harvard University Press, 1955.

Davis, Keith F. " 'A Terrible Distinctiveness': Photography of the Civil War Era." In Martha A. Sandweiss, ed., *Photography in Nineteenth-Century America*. New York: Harry N. Abrams, 1991.

Dean, Eric, T. *Shook over Hell: Post-traumatic Stress, Vietnam, and the Civil War*. Cambridge, MA: Harvard University Press, 1997.

Derickson, Alan. *Health Security for All: Dreams of Universal Health Care in America*. Baltimore: Johns Hopkins University Press, 2005.

Deville, Kenneth Allen. *Medical Malpractice in Nineteenth Century America*. New York: New York University Press, 1990.

Dickson, Paul and Thomas B. Allen. *The Bonus Army: An American Epic*. New York: Walker & Co., 2005.

Dodd, Wynelle and Donald. *Historical Statistics of the United States, 1790 to 1970*. Birmingham: University of Alabama Press, 1973.

Dowling, Timothy, ed. *Personal Perspectives: World War I*. Santa Barbara, CA: ABC-CLIO, 2006.

Dunlevy, Crystal Leigh. "The Contributions of Women of the National Society for the Promotion of Industrial Education to the Development of Vocational Education for Women, 1906–1917." PhD diss., Rutgers University, 1988.

Dwyer, Ellen. "Psychiatry and Race during World War II." *Journal of the History of Medicine and Allied Sciences* 61, no. 2 (2006): 117–43.

Eghigian, Greg. *Making Security Social: Disability, Insurance, and the Birth of the Social Entitlement State in Germany*. Ann Arbor: University of Michigan Press, 2000.

Ellis, Mark. *Race, War, and Surveillance: African Americans and the United States Government during World War I*. Bloomington: Indiana University Press, 2001.

Engel, Jonathan. *Doctors and Reformers: Discussion and Debate over Health Policy, 1925–1950*. Columbia: University of South Carolina Press, 2002.

English, Peter C. "Not Miniature Men and Women." In Loretta M. Kopelman and John C. Moskop, eds., *Children and Health Care: Moral and Social Issues*. Dordrecht: Kluwer, 1989.

Espinosa, Mariola. *Epidemic Invasions: Yellow Fever and the Limits of Cuban Independence*. Chicago: University of Chicago Press, 2009.

Fairchild, Amy. *Science at the Borders: Immigrant Medical Inspection and the Shaping of the Modern Industrial Labor Force.* Baltimore: Johns Hopkins University Press, 2003.

Figg, Laurann and Jane Farrell-Beck. "Amputation in the Civil War: Physical and Social Dimensions." *Journal of the History of Medicine and Allied Sciences* 48 (1993): 454–75.

Flanagan, Maureen. *Seeing with their Hearts: Chicago Women and the Vision of the Good City, 1871–1933.* Princeton: Princeton University Press, 2002.

Foucault, Michel. *Discipline and Punish: The Birth of the Prison.* Trans. Alan Sheridan. 2nd ed. New York: Vintage Books, 1995.

Fox, Daniel. *Power and Illness: The Failure and Future of American Health Policy.* Berkeley: University of California Press, 1993.

Fox, Daniel M., and Christopher Lawrence. *Photographing Medicine: Images and Power in Britain and America since 1840.* New York: Greenwood Press, 1988.

Fox, Renée. "The Medicalization and Demedicalization of American Society." *Daedalus* 106 (Winter 1977): 9–22.

Fraser, Nancy, and Linda Gordon, "A Genealogy of Dependency: Tracing a Keyword of the U.S. Welfare State." *Signs: Journal of Women in Culture and Society* 19 (Winter 1994): 309–36.

Gamble, Vanessa Northington. *Making a Place for Ourselves: The Black Hospital Movement, 1920–1945.* New York: Oxford University Press, 1995.

Gallagher, Hugh G. *FDR's Splendid Deception.* New York: Dodd, Mead, 1985.

Gariepy, Thomas P. "The Introduction and Acceptance of Listerian Antisepsis in the United States." *Journal of the History of Medicine and Allied Sciences* 49 (1994): 167–206.

Gavin, Lettie. *American Women in World War I: They Also Served.* Niwot: University of Colorado Press, 1997.

Gaughan, A. "Woodrow Wilson and the Legacy of the Civil War." *Civil War History* 43, no. 3 (1997): 225–42.

Geison, Gerald. *The Private Science of Louis Pasteur.* Princeton: Princeton University Press, 1995.

Gelber, Scott. "A 'Hard-Boiled Order': The Reeducation of Disabled World War I Veterans in New York City." *Journal of Social History* 39, no. 1 (2005): 161–80.

Gerber, David A. "Disabled Veterans, the State, and the Experience of Disability in Western Societies, 1914–1950." *Journal of Social History* 36, no. 4 (Summer 2003): 899–916

Gerber, David A., ed. *Disabled Veterans in History.* Ann Arbor: University of Michigan Press, 2000.

Gerstle, Gary. *Working-Class Americanism: The Politics of Labor in a Textile City, 1914–1960.* 2nd ed. Princeton: Princeton University Press, 2002.

Giesberg, Judith Ann. *Civil War Sisterhood: The U.S. Sanitary Commission and Women's Politics in Transition.* Boston: Northeastern University Press, 2000.

Gillespie, Richard. "Industrial Fatigue and the Discipline of Physiology." In Gerald L. Geison, ed., *Physiology in the American Context*. Bethesda: American Physiological Society, 1987.

Gillett, Mary C. *The Army Medical Department, 1865–1917*. Washington, DC: Center of Military History, United States Army, 1995.

Gilmore, Glenda E. *Gender and Jim Crow: Women and the Politics of White Supremacy in North Carolina, 1896–1920*. Chapel Hill: University of North Carolina Press, 1996.

Ginn, Richard V. N. *The History of the U.S. Army Medical Service Corps*. Washington, DC: Office of the Surgeon General and Center of Military History, United States Army, 1997.

Girard, Marion. *A Strange and Formidable Weapon: British Responses to World War I Poison Gas*. Lincoln: University of Nebraska Press, 2008.

Glickman, Lawrence B. *A Living Wage: American Workers and the Making of Consumer Society*. Ithaca, NY: Cornell University Press, 1997.

Golden, Janet, Richard A. Meckel, and Heather Munro Prescott, ed. *Children and Youth in Sickness and in Health: A Historical Handbook and Guide*. Westport, CT: Greenwood Press, 2004.

Gordon, Colin. *Dead on Arrival: The Politics of Health Care in Twentieth Century America*. Princeton: Princeton University Press, 2003.

Gordon, Linda. *Pitied but Not Entitled: Single Mothers and the History of Welfare, 1890–1935*. New York: Free Press, 1994.

———. *The Great Arizona Orphan Abduction*. Cambridge, MA: Harvard University Press, 1999.

Gordon, Lynn D. *Gender and Higher Education in the Progressive Era*. New Haven: Yale University Press, 1990.

Gordon, Howard R. D. *The History and Growth of Career and Technical Education in America*, 3rd ed. Long Grove, IL: Waveland Press, 2007.

Grantham, Dewey W. Jr. *Hoke Smith and the Politics of the New South*. Baton Rouge: Louisiana State University Press, 1958.

———. *The Regional Imagination: The South and Recent American History*. Nashville: Vanderbilt University Press, 1979.

Gritzer, Glenn and Arnold Arluke. *The Making of Rehabilitation: A Political Economy of Medical Specialization, 1890–1980*. Berkeley: University of California Press, 1985.

Grob, Gerald N. *From Asylum to Community: Mental Health Policy in Modern America*. Princeton, NJ: Princeton University Press, 1991.

———. *The Mad among Us: A History of the Care of America's Mentally Ill*. New York: Free Press, 1994.

Guy, Chester C. "The Railroad Surgeon, His Association, His Hospitals, His Past, His Present." *Industrial Medicine and Surgery* 32 (1963): 351–60.

Guyatt, Mary. "Better Legs: Artificial Limbs for British Veterans of the First World War," *Journal of Design History* 14 (2001): 307–25.

Haber, Carole. *Beyond Sixty-Five: The Dilemma of Old Age in America's Past.* New York: Cambridge University Press, 1983.

Haber, L. F. *The Poisonous Cloud: Chemical Warfare in the First World War.* New York: Oxford University Press, 1985.

Haines, Michael R. "Fertility and Mortality, by Race: 1800–2000." In Susan B. Carter, ed., *Historical Statistics of the United States,* vol. 1. New York: Cambridge University Press, 2006.

Haller, Beth and Robin Larsen. "Persuading Sanity: Magic Lantern Images and the Nineteenth-Century Moral Treatment in America." *Journal of American Culture* 28, no. 3 (2005): 259–72.

Hallett, Christine. "The Personal Writings of First World War Nurses: A Study of the Interplay of Authorial Intention and Scholarly Interpretation." *Nursing Inquiry* 14 (2007): 320–29.

Halpern, Sydney. *American Pediatrics: The Social Dynamics of Professionalism, 1880–1980.* Berkeley and Los Angeles: University of California Press, 1988.

Harries, Meirion, and Susie Harries. *The Last Days of Innocence: America at War, 1917–1918.* New York: Random House, 1997.

Harrison, Mark. "Disease, Discipline and Dissent: The Indian Army in France and England, 1914–1915." In Roger Cooter, Mark Harrison, and Steve Sturdy, eds., *Medicine and Modern Warfare,* 185–204. Amsterdam: Rodopi, 1999.

Hays, Samuel P. *Conservation and the Gospel of Efficiency.* Cambridge, MA: Harvard University Press, 1959.

Heap, Ruby. " 'Salvaging War's Waste': The University of Toronto and the 'Physical Reconstruction' of Disabled Soldiers during the First World War." In Edgar-André Montigny and Lori Chambers, eds., *Ontario since Confederation: A Reader,* 214–34. Toronto: University of Toronto Press, 2000.

Heller, Jonathan. "Photographing the Great War." *Prologue: Quarterly of the National Archives and Record Administration* 30, no. 3 (Fall 1998): 220–27.

Henderson, Richard B. *Maury Maverick: A Political Biography.* Austin: University of Texas Press, 1970.

Henry, Robert S. *The Armed Forces Institute of Pathology: Its First Century, 1862–1962.* Washington, DC: Office of the Surgeon General, 1964.

Herschbach, Lisa Marie. "Prosthetic Reconstructions: Making the Industry, Remaking the Body, Modelling the Nation." *History Workshop Journal* 44 (Autumn 1997): 28–33.

———. "Fragmentation and Reunion: Medicine, Memory, and the Body in the American Civil War." PhD diss., Harvard University, 1997.

Huebner, Andrew J. *The Warrior Image: Soldiers in American Culture from the Second World War to the Vietnam Era.* Chapel Hill: University of North Carolina Press, 2008.

Hickel, Karl Walter. "Entitling Citizens: World War I, Progressivism, and the Origins of the American Welfare State, 1917–1928." PhD diss., Columbia University, 1999.

———. "War, Region, and Social Welfare: Federal Aid to Servicemen's Dependents in the South, 1917–1921." *Journal of American History* 87, no. 4 (March 2001): 1362–91.

Higonnet, Margaret Randolph, et al., eds. *Behind the Lines: Gender and the Two World Wars*. New Haven: Yale University Press, 1987.

Hirsch, Edwin F. *Frank Billings: A Leader in Chicago Medicine*. Chicago: University of Chicago Press, 1966.

Hodges, Patricia A. M. "Perspectives on History: Military Dietetics in Europe during World War I." *Journal of the American Dietetic Association* 93 (August 1993): 897–901.

Hoffman, Beatrix. *The Wages of Sickness: The Politics of Health Insurance in Progressive America*. Chapel Hill: University of North Carolina Press, 2001.

Hoganson, Kristin L. *Fighting for American Manhood: How Gender Politics Provoked the Spanish-American and Philippine-American Wars*. Yale Historical Publications. New Haven: Yale University Press, 1998.

Holcombe, R. G. "Veterans Interests and the Transition to Government Growth: 1870–1915." *Public Choice* 99, no. 3 (1999): 311–26.

Holt, Stull W. *The Federal Board for Vocational Education: Its History, Activities, and Organization*. New York: Appleton, 1922.

Hounshell, David A. *From the American System to Mass Production, 1800–1932: The Development of Manufacturing Technology in the United States*. Baltimore: Johns Hopkins University Press, 1984.

Howell, Joel D. *Technology in the Hospital: Transforming Patient Care in the Early Twentieth Century*. Baltimore: Johns Hopkins University Press, 1995

Hutchinson, John F. *Champions of Charity: War and the Rise of the Red Cross*. Boulder, CO: Westview Press, 1996.

Illich, Ivan. *Medical Nemesis*. London: Marion Boyars, 1976.

Jarvis, Christina S. *The Male Body at War: American Masculinity during World War II*. DeKalb: Northern Illinois University Press, 2004.

Jefferson, Robert F. " 'Enabled Courage': Disability, and Black World War II Veterans in Postwar America." *Historian* 65 (September 2003): 1102–24.

Jeffords, Susan. *The Remasculinization of America: Gender and the Vietnam War*. Bloomington: Indiana University Press, 1989.

Johnson, Donald. "Wilson, Burleson, and Censorship in the First World War." *Journal of Southern History* 28 (February 1962): 46–58.

K., J.G. "Joel Ernest Goldthwait, 1866–1961." *Journal of Bone and Joint Surgery* 43 (1961): 463–64.

Kaldor, Mary. *The Baroque Arsenal*. New York: Hill and Wang, 1981.

Kantor, Harvey. "Work, Education, and Vocational Reform: The Ideological Origins of Vocational Education, 1890–1920." *American Journal of Education* 94, no. 4 (August 1986): 401–26.

Kantor, Harvey, and David B. Tyack, eds., *Work, Youth, and Schooling: Historical Perspectives on Vocationalism in American Education*. Stanford: Stanford University Press, 1982.

Kasson, John F. *Houdini, Tarzan, and the Perfect Man: The White Male Body and the Challenge of Modernity in America.* New York: Hill and Wang, 2001.

Katz, Michael B. *In the Shadow of the Poorhouse: A Social History of Welfare.* New York: Basic Books, 1986.

———. *Reconstructing American Education.* Cambridge, MA: Harvard University Press, 1987.

Katznelson, Ira. *Schooling for All: Class, Race, and the Decline of the Democratic Ideal.* New York: Basic Books, 1985.

Keene, Jennifer. *Doughboys, the Great War, and the Remaking of America.* Baltimore: Johns Hopkins University Press, 2001.

———. "Protest and Disability: A New Look at African American Soldiers during the First World War." In Pierre Purseigle, ed., *Warfare and Belligerence: Perspectives in First World War Studies*, 215–41. Leiden: Brill Press, 2005.

———. *World War I.* Westport, CT: Greenwood Press, 2006.

Keers, R. Y. *Pulmonary Tuberculosis: A Journey down the Centuries.* London: Bailliere Tindall, 1978.

Kelly, Patrick J. *Creating a National Home: Building the Veterans' Welfare State, 1860–1900.* Cambridge, MA: Harvard University, 1997.

———. "The Election of 1896 and the Restructuring of Civil War Memory." *Civil War History* 49, no. 3 (2003): 254–82.

Kennedy, David M. *Over Here: The First World War and American Society.* New York: Oxford University Press, 1980.

Kevles, Daniel J. "Testing the Army's Intelligence: Psychologists and the Military in World War I." *Journal of American History* 55 (1968): 565–81.

———. *In the Name of Eugenics: Genetics and the Uses of Human Heredity.* New York: Knopf, 1985.

Klein, Jennifer. *For All These Rights: Business, Labor, and the Shaping of America's Public-Private Welfare State.* Princeton, NJ: Princeton University Press, 2003.

Kopelman, Loretta M., and John C. Moskop, eds. *Children and Health Care: Moral and Social Issues.* Dordrecht: Kluwer, 1989.

Koven, Seth. "Remembering and Dismemberment: Crippled Children, Wounded Soldiers, and the Great War in Great Britain." *American Historical Review* 99 (October 1994): 1167–1202.

Koven, Seth, and Sonya Michel. "Womanly Duties: Maternalist Politics and the Origins of Welfare States in France, Germany, Great Britain, and the United States, 1880–1920." *American Historical Review* 95 (October 1990): 1076–1108.

Kudlick, Catherine J. "Why We Need Another 'Other.'" *American Historical Review* 108 (June 2003): 763–93.

Ladd-Taylor, Molly. *Mother-Work: Women, Child Welfare, and the State, 1890–1930.* Urbana: University of Illinois Press, 1994.

Ladd-Taylor, Molly, and Laurie Umansky, eds. *"Bad" Mothers: The Politics of Blame in Twentieth-Century America.* New York: New York University Press, 1998.

Lansing, Michael J. " 'Salvaging the Man Power of America': Conservation, Manhood, and Disabled Veterans during World War I." *Environmental History* 14 (January 2009): 32–57.

Larkin, Gerald. *Occupational Monopoly and Modern Medicine.* London: Tavistock Publications, 1983.

Larson, Elaine. "Innovation in Health Care: Antisepsis as a Case Study." *American Journal of Public Health* 89 (January 1999): 92–99

Larsson, Marina. *Shattered Anzacs: Living with the Scars of War.* Sydney, Australia: University of New South Wales Press, 2009.

Lawrence, Christopher, ed. *Medical Theory, Surgical Practice: Studies in the History of Surgery.* London: Routledge, 1992.

Lears, T. J. Jackson. *Fables of Abundance: A Cultural History of Advertising in America.* New York: Basic Books, 1994.

———. *No Place of Grace: Antimodernism and the Transformation of American Culture, 1880–1920.* 2nd ed. Chicago: University of Chicago Press, 1994.

———. *Rebirth of a Nation: The Making of Modern America, 1877–1920.* New York: HarperCollins, 2009.

Leavitt, Judith Walzer. *Brought to Bed: Childbearing in America, 1750 to 1950.* New York: Oxford University Press, 1986.

Lederer, Susan E. "Hideyo Noguchi's Leutin Experiment and Antivivisectionists." *Isis* 76, no. 1 (March 1985): 31–48.

———. *Subjected to Science: Human Experimentation in America before the Second World War.* Baltimore: Johns Hopkins University Press, 1995.

———. "Repellent Subjects: Hollywood Censorship and Surgical Images in the 1930s." *Literature and Medicine* 17, no. 1 (1998): 91–113.

———. *Flesh and Blood: Organ Transplantation and Blood Transfusion in Twentieth Century America.* Oxford and New York: Oxford University Press, 2008.

Leese, Peter. *Shell Shock: Traumatic Neurosis and the British Soldiers of the First World War.* Hampshire and New York: Palgrave Macmillan, 2002.

Lentz-Smith, Adriane. *Freedom Struggles: African Americans and World War I.* Cambridge, MA: Harvard University Press, 2009.

Lerner, Barron H. *Contagion and Confinement: Controlling Tuberculosis along the Skid Road.* Baltimore: Johns Hopkins University Press, 1998.

Lerner, Paul Frederick. *Hysterical Men: War, Psychiatry, and the Politics of Trauma in Germany, 1890–1930.* Ithaca, NY: Cornell University Press, 2003.

Lindenmeyer, Kriste. *Right to Childhood: The U.S. Children's Bureau and Child Welfare, 1912–46.* Urbana: University of Illinois Press, 1997.

Linker, Beth. "The Business of Ethics: Gender, Medicine, and the Professional Codification of the American Physiotherapy Association, 1918–1935." *Journal of the History of Medicine and Allied Sciences* 60 (July 2005): 321–54.

———. "Strength and Science: Gender, Physiotherapy, and Medicine in Early Twentieth Century America." *Journal of Women's History* 17, no. 3 (Fall 2005): 105–32.

———. "Feet for Fighting: Locating Disability and Social Medicine in World War I America." *Social History of Medicine* 20 (2007): 91–109.

———. "Rest Cure." In Susan Burch, ed., *Encyclopedia of American Disability History* 3: 780–81. New York: Infobase Publishing, 2009.

Lord, Alexandra M. "'Naturally Clean and Wholesome': Women, Sex Education, and the United States Public Health Service, 1918–1928." *Social History of Medicine* 17, no. 3 (December 1, 2004): 423–41.

Logue, Larry M. "Union Veterans and Their Government: The Effects of Public Policies on Private Lives." *Journal of Interdisciplinary History* 22, no. 3 (Winter 1992): 411–34.

Logue, Larry M., and Peter Blanck. "'Benefit of the Doubt': African-American Civil War Veterans and Pensions." *Journal of Interdisciplinary History* 38, no. 3 (2008): 377–99.

Logue, Larry M., and Peter Blanck. *Race, Ethnicity, and Disability: Veterans and Benefits in Post–Civil War America.* Cambridge: Cambridge University Press, 2010.

Long, Diana E., and Janet Golden, eds. *The American General Hospital: Communities and Social Contexts.* Ithaca, NY: Cornell University Press, 1989.

Longmore, Paul K., and Lauri Umansky, eds. *The New Disability History: American Perspectives.* New York: New York University Press, 2001.

Longmore, Paul K. *Why I Burned My Book and Other Essays on Disability.* Philadelphia: Temple University Press, 2003.

Lunbeck, Elizabeth. *The Psychiatric Persuasion: Knowledge, Gender, and Power in Modern America.* Princeton, NJ: Princeton University Press, 1994.

Manchester, Katherine E., and Helen B. Gearin. "Dieticians before World War II." In Robert S. Anderson, Harriet S. Lee, and Myra L. McDaniel, eds., *Army Medical Specialist Corps,* 15–39. Washington, DC: Office of the Surgeon General, 1968.

Mangan, J. A., and Roberta J. Park, eds. *From 'Fair Sex' to Feminism: Sport and the Socialization of Women in the Industrial and Post-Industrial Eras.* London: Frank Cass & Co., 1987.

Mankin, Henry J. "Boston's Contributions to the Development of Orthopaedics in the United States." *Clinical Orthopaedics and Related Research* 458 (2000): 47–54.

Marble, Sanders. *Rehabilitating the Wounded: Historical Perspective on Army Policy.* Washington, DC: Office of Medical History, 2008.

Maroney, Mildred. "Veterans' Benefits." In Lewis Meriam and Karl T. Schlotterbeck, eds., *Cost and Financing of Social Security,* 96–129. Washington, DC: Brookings Institute, 1950.

McBride, David. *From TB to AIDS: Epidemics among Urban Blacks since 1900.* Albany: State University of New York Press, 1991.

McCormick, Virginia E. "The Talented Sherwoods: Poets and Politicians." *Northwest Ohio Quarterly* 52, no. 3 (1980): 244–53.

McClure, Arthur F., James Riley Chrisman, and Perry Mock, eds. *Education for Work: The Historical Evolution of Vocational and Distributive Education in America.* Rutherford, NJ: Fairleigh Dickinson University Press, 1985.

Meckel, Richard A. *Save the Babies: American Public Health Reform and the Prevention of Infant Mortality, 1850–1929.* Baltimore: Johns Hopkins University Press, 1990.

———."Open Air Schools and the Tuberculous Child in Early Twentieth Century America." *Archives of Pediatric and Adolescent Medicine* 150 (1996): 91–96.

Mendelsohn, Everett. "Science, Scientists, and the Military." In John Krige and Dominique Pestre, eds., *Companion to Science in the Twentieth Century.* 2nd ed. London: Routledge, 2003.

Micale, Mark S. *Hysterical Men: The Hidden History of Male Nervous Illness.* Cambridge, MA: Harvard University Press, 2008.

Michel, Sonya. "The Limits of Maternalism: Policies toward American Wage-Earning Mothers during the Progressive Era." In Seth Koven and Sonya Michel, eds., *Mothers of a New World: Maternalist Politics and the Origins of Welfare States.* New York: Routledge, 1993.

Milkman, Ruth. *Gender at Work: The Dynamics of Job Segregation by Sex during World War II.* Urbana: University of Illinois Press, 1987.

Morantz-Sanchez, Regina. *Sympathy and Science: Women Physicians in American Medicine.* Oxford: Oxford University Press, 1985.

More, Ellen Singer, and Maureen A. Milligan, eds. *The Empathic Practitioner: Empathy, Gender, and Medicine.* New Brunswick: Rutgers University Press, 1994.

Mörgeli, Christoph. *The Surgeon's Stage: A History of the Operating Room.* Basel: Roche, 1999.

Morton, Desmond and Glenn Wright. *Winning the Second Battle: Canadian Veterans and the Return to Civilian Life, 1915–1930.* Toronto: University of Toronto Press, 1987.

Mosse, George L. *Fallen Soldiers: Reshaping the Memory of the World Wars.* Oxford: Oxford University Press, 1990.

Muncy, Robyn. *Creating a Female Dominion in American Reform, 1890–1935.* Oxford: Oxford University Press, 1991.

Murphy, Wendy. *Healing the Generations: A History of Physical Therapy and the American Physical Therapy Association.* Alexandria: American Physical Therapy Association, 1995.

Nielsen, Kim E. *The Radical Lives of Helen Keller.* New York: New York University Press, 2004.

Nicholls, D. A., and J. Cheek. "Physiotherapy and the Shadow of Prostitution: The Society of Trained Masseuses and the Massage Scandals of 1894." *Social Science and Medicine* 62, no. 9 (2006): 2336–48.

Nicholson, Jesse T. "Founders of American Orthopedics." *Journal of Bone and Joint Surgery of America* 481 (1966): 582–97.

Numbers, Ronald L. *Almost Persuaded: American Physicians and Compulsory Health Insurance, 1912–1920*. Baltimore: Johns Hopkins University Press, 1978.

Oberman, C. Esco. *A History of Vocational Rehabilitation in America*. Minneapolis: T. S. Denison and Company, Inc., 1965.

Orr, H. Winnett. *Fifty Years of the American Orthopedic Association*. Lincoln Nebraska, 1937.

Ortiz, Stephen R. "The 'New Deal' for Veterans: The Economy Act, the Veterans of Foreign Wars, and the Origins of New Deal Dissent." *Journal of Military History* 70 (2006): 415–38.

———. *Beyond the Bonus March and GI Bill: How Veteran Politics Shaped the New Deal Era*. New York: New York University Press, 2010.

Oshinsky, David M. *Polio: An American Story*. Oxford: Oxford University Press, 2005.

Ott, Katherine. *Fevered Lives: Tuberculosis in American Culture since 1870*. Cambridge, MA: Harvard University Press, 1996.

Ott, Katherine, David Serlin, and Stephen Mihm, eds. *Artificial Parts, Practical Lives: Modern Histories of Prosthetics*. New York: New York University Press, 2002.

Panchasi, Roxanne. "Reconstructions: Prosthetics and the Rehabilitation of the Male Body in World War I France." *Differences: A Journal of Feminist Cultural Studies* 7, no. 3 (1995): 109–40.

Pernick, Martin. *A Calculus of Suffering: Pain, Professionalism, and Anesthesia in Nineteenth-Century America*. New York: Columbia University Press, 1985.

———. *The Black Stork: Eugenics and the Death of "Defective" Babies in American Medicine and Motion Pictures since 1915*. New York: Oxford University Press, 1996.

Perry, Heather R. "Recycling the Disabled: Army, Medicine, and Society in World War I Germany." PhD diss., Indiana University, 2005.

———. "Re-Arming the Disabled Veteran: Artificially Rebuilding State and Society in World War I Germany." In Katherine Ott, David Serlin, and Stephen Mihm, eds., *Artificial Parts, Practical Lives: Modern Histories of Prosthetics*, 75–101. New York: New York University Press, 2002.

Peterson, Theodore Bernard. *Magazines in the Twentieth Century*. Urbana: University of Illinois Press, 1964.

Pettegrew, J. " 'The Soldier's Faith': Turn-of-the-century Memory of the Civil War and the Emergence of Modern American Nationalism." *Journal of Contemporary History* 31, no. 1 (1996): 49–73.

Pierce, John R. *Yellow Jack: How Yellow Fever Ravaged America and Walter Reed Discovered Its Deadly Secrets*. Hoboken, NJ: J. Wiley, 2005.

Pope, Steven W. "An Army of Athletes: Playing Field, Battlefields, and the American Military Sporting Experience, 1890–1920." *Journal of Military History* 59, no. 3 (July 1995): 435–56.

Porter, Roy. "The Patient's View: Doing Medical History from Below," *Theory and Society* 14 (1985): 167–74.

Purtilo, Ruth. "The American Physical Therapy Association's Code of Ethics: Its Historical Foundations." *Physical Therapy* 57 (September 1977): 1001–6.

Quiroga, Virginia A. M. "Female Lay Managers and Scientific Pediatrics at Nursery and Child's Hospital, 1854–1910." *Bulletin of the History of Medicine* 60 (1986): 192–208.

———. *Occupational Therapy: The First Thirty Years, 1900–1930.* Bethesda, MD: American Occupational Therapy Association, 1995.

Rabinbach, Anson. *The Human Motor: Energy, Fatigue, and the Origins of Modernity.* Berkeley: University of California Press, 1990.

Resch, John P. "Politics and Public Culture: The Revolutionary War Pension Act of 1818." *Journal of the Early Republic* 8 (Summer 1988): 139–58.

———. *Suffering Soldiers: Revolutionary War Veterans, Moral Sentiment, and Political Culture in the Early Republic.* Amherst: University of Massachusetts Press, 1999.

Reverby, Susan. *Ordered to Care: The Dilemma of American Nursing, 1850–1945.* Cambridge: Cambridge University Press, 1987.

Reznick, Jeffrey S. *Healing the Nation: Soldiers and the Culture of Caregiving in Britain during the Great War.* Manchester: Manchester University Press, 2004.

Rhode, Michael. "Photography and the Army Medical Museum, 1862–1945." *Architext* 4, no. 2 (March 1995): 7–10.

Rodgers, Daniel T. *The Work Ethic in Industrial America, 1850–1920.* Chicago: University of Chicago Press, 1978.

———. *Atlantic Crossings: Social Politics in a Progressive Age.* Cambridge, MA: Belknap Press of Harvard University Press, 1998.

Rodgers, Daniel T., and David B. Tyack, "Work, Youth, and Schooling: Mapping Critical Research Areas." In Harvey Kantor and David B. Tyack, eds., *Work, Youth, and Schooling: Historical Perspectives on Vocationalism in American Education.* Stanford: Stanford University Press, 1982.

Rogers, Naomi. *Dirt and Disease: Polio before FDR.* New Brunswick, NJ: Rutgers University Press, 1992.

Romano, Sally D. "The Dark Side of the Sun: Skin Cancer, Sunscreen, and Risk in Twentieth-Century America." PhD diss., Yale University, 2006.

Rose, Sarah. "'Crippled' Hands: Disability in Labor and Working-Class History." *Labor: Studies in Working-Class History of the Americas* 2 (2005): 27–54.

Rosenberg, Charles E. *The Care of Strangers: The Rise of America's Hospital System.* New York: Basic Books, 1987.

———. "Representing Medicine: Philadelphia, Health, and Photography, 1860–1945." In Janet Golden and Charles E. Rosenberg, eds., *Pictures of Health: A Photographic History of Health Care in Philadelphia, 1860–1945.* Philadelphia: University of Pennsylvania Press, 1991.

Rosenburg, R. B. "'Empty Sleeves and Wooden Pegs': Disabled Confederate Veterans in Image and Reality," in David A. Gerber, ed., *Disabled Veterans in History,* 204–28. Ann Arbor: University of Michigan Press, 2000.

Rosenzweig, Roy. *Eight Hours for What We Will: Workers and Leisure in an Industrial City, 1870–1920.* Cambridge: Cambridge University Press, 1983.

Rosner, David. *A Once Charitable Enterprise: Hospitals and Health Care in Brooklyn and New York, 1885–1915.* Cambridge and New York: Cambridge University Press, 1982.

Rosner, David, and Gerald Markowitz. "The Early Movement for Occupational Safety and Health, 1900–1917." In Judith Walzer Leavitt and Ronald L. Numbers, eds., *Sickness and Health in America: Readings in the History of Medicine and Public Health,* 467–82. 3rd ed. Madison: University of Wisconsin Press, 1997.

Rosner, David, and Gerald Markowitz, eds. *Dying for Work: Workers' Safety and Health in Twentieth-Century America.* Bloomington: Indiana University Press, 1987.

Rothman, David J. *The Discovery of the Asylum: Social Order and Disorder in the New Republic.* Boston: Little, Brown, 1971.

Rothman, Sheila M. *Living in the Shadow of Death: Tuberculosis and the Social Experience of Illness in American History.* New York: Basic Books, 1994.

Rutkow, Ira. "A Selective History of Hernia Surgery in the Late Eighteenth Century: The Treatises of Percivall Pott, Jean Louis Petit, D. August Gottlieb Richter, Don Antonio de Gimbernat, and Pieter Camper." *Surgical Clinics of North America* 83, no. 5 (October 2003): 1021–44.

Shaffer, Donald R. " ' I do not suppose that Uncle Sam looks at the skin': African Americans and the Civil War Pension System, 1865–1934." *Civil War History* 46, no. 2 (June 2000): 132–47.

Schaffer, Ronald. *America in the Great War: The Rise of the War Welfare State.* New York: Oxford University Press, 1991.

Schalick, Walton O. "Children, Disability, and Rehabilitation in History." *Pediatric Rehabilitation* 4 (2001): 91–95.

Schlich, Thomas. *Surgery, Science, and Industry: A Revolution in Fracture Care, 1950s–1990s.* Houndmills, Basingstoke: Palgrave, 2002.

———. "The Emergence of Modern Surgery." In Deborah Brunton, ed., *Medicine Transformed: Health, Disease, and Society in Europe, 1800–1930,* 61–91. Manchester: Manchester University Press, 2004.

———. "Trauma Surgery and Traffic Policy in Germany in the 1930s: A Case Study in the Coevolution of Modern Surgery and Society." *Bulletin of the History of Medicine* 80, no. 1 (2006): 73–94.

———. "The Perfect Machine: Lorenz Böhler's Rationalized Fracture Treatment in WWI." *Isis* 100 (2009): 758–91.

Scotch, Richard. *From Good Will to Civil Rights: Transforming Disability Policy.* 2nd ed. Philadelphia: Temple University Press, 2001.

Scott, Emmett J. *The American Negro in the World War.* 2nd ed. New York: Arno Press, 1969.

Schwartz, Hillel. "Torque: The New Kinaesthetic of the Twentieth Century." In Jonathan Crary and Sanford Kwinter, eds., *Incorporations.* New York: Zone, 1992.

Schweik, Susan M. *The Ugly Laws: Disability in Public.* New York: New York University, 2009.

Selcer, Perrin. "Standardizing Wounds: Alexis Carrel and the Scientific Management of Life in the First World War." *British Journal for the History of Science* 41 (2008): 73–107.

Serlin, David. "Crippling Masculinity: Queerness and Disability in U.S. Military Culture, 1800–1945." *GLQ: Journal of Lesbian and Gay Studies* 9 (2003): 149–79.

———. *Replaceable You: Engineering the Body in Postwar America.* Chicago: University of Chicago Press, 2004.

Shands, Alfred R. "DeForest Willard, Philadelphia's Pioneer Orthopaedic Surgeon (1846–1910)." *Current Practice in Orthopedic Surgery* 4 (1969): 43–57.

Sigerist, Henry E. *Medicine and Human Welfare.* New Haven: Yale University Press, 1941.

Silver, Roberta. "An Analysis of Charles Allen Prosser's Conception of Secondary Education in the United States." PhD diss., Loyola University of Chicago, 1991.

Skocpol, Theda. *Protecting Soldiers and Mothers: The Political Origins of Social Policy in the United States.* Cambridge: Harvard University Press, 1992.

Slavishak, Edward. "Artificial Limbs and Industrial Workers' Bodies in Turn-of-the-Century Pittsburgh." *Journal of Social History* 37, no. 2 (2003): 365–88.

Smith, Merritt Roe. *Harpers Ferry Armory and the New Technology: The Challenge of Change.* Ithaca, NY: Cornell University Press, 1977.

Smith, Steven J., R. Mark Evans, Micaela Sullivan-Fowler, and William R. Hendee. "Use of Animals in Biomedical Research: Historical Role of the American Medical Association and the American Physician." *Archives of Internal Medicine* 148 (August 1988): 1849–53.

Smith-Rosenberg, Caroll. *Disorderly Conduct: Visions of Gender in Victorian America.* Oxford: Oxford University Press, 1985.

Smith-Rosenberg, Caroll, and Charles Rosenberg. "The Female Animal: Medical and Biological Views of Women and Their Role in Nineteenth-century America." In J. A. Mangan and Roberta J. Park, eds., *From 'Fair Sex' to Feminism: Sport and the Socialization of Women in the Industrial and Post-Industrial Eras,* 13–37. London: Frank Cass, 1987.

Snyder, Sharon and David T. Mitchell. *Cultural Locations of Disability.* Chicago: University of Chicago Press, 2006.

Standlee, Mary W. *Borden's Dream: The Walter Reed Army Medical Center in Washington, D.C.* Washington, DC: Borden Institute, 2009.

Starr, Paul. *The Social Transformation of American Medicine.* New York: Basic Books, 1982.

———. *The Discarded Army: Veterans after Vietnam, The Nader Report on Vietnam Veterans and the Veterans Administration.* New York: Charterhouse, 1973.

Stern, Alexandra Minna, and Howard Markel, eds. *Formative Years: Children's Health in the United States, 1880–2000.* Ann Arbor: University of Michigan Press, 2002.

Stevens, Rosemary. *American Medicine and the Public Interest.* 2nd ed. Berkeley: University of California Press, 1988.

———. "Can the Government Govern? Lessons from the Formation of the Veterans Administration." *Journal of Health Politics, Policy, and Law* 16.2 (Summer 1991): 281–305.

———. *In Sickness and in Wealth: American Hospitals in the Twentieth Century.* 2nd ed. Baltimore: Johns Hopkins University Press, 1999.

Stone, Deborah. *The Disabled State.* Philadelphia: Temple University Press, 1984.

Stradling, David, ed. *Conservation in the Progressive Era: Classic Texts.* Seattle: University of Washington Press, 2004.

Striker, Henri-Jacques. *A History of Disability.* Translated by William Sayers. Ann Arbor: University of Michigan Press, 1999.

Summers, Mark W. *Party Games: Getting, Keeping, and Using Power in Gilded Age Politics.* Chapel Hill: University of North Carolina Press, 2004.

Szasz, Thomas. *The Manufacture of Madness: A Comparative Study of the Inquisition and the Mental Health Movement.* New York: Harper Row, 1970.

Teller, Michael E. *The Tuberculosis Movement: A Public Health Campaign in the Progressive Era.* New York: Greenwood Press, 1988.

Thompson, John A. *Reformers and War: American Progressive Publicists and the First World War.* Cambridge: Cambridge University Press, 1987.

Thompson, John D. and Grace Goldin. *The Hospital: A Social and Architectural History.* New Haven: Yale University Press, 1975.

Thomson, Rosemarie Garland, ed. *Freakery: Cultural Spectacles of the Extraordinary Body.* New York: New York University Press, 1996.

Tomes, Nancy. *A Generous Confidence: Thomas Story Kirkbride and the Art of Asylum-Keeping, 1840–1883.* Cambridge: Cambridge University Press, 1984.

———. *The Gospel of Germs: Men, Women, and the Microbe in American Life.* Cambridge, MA: Harvard University Press, 1998.

Tracy, Sarah W. *Alcoholism in America: From Reconstruction to Prohibition.* Baltimore: Johns Hopkins University Press, 2005.

Venzon, Anne Cipriano, ed. *The United States in the First World War: An Encyclopedia.* New York: Garland Publishing, 1995.

Verbrugge, Martha. *Able-Bodied Womanhood: Personal Health and Social Change in Nineteenth-Century Boston.* Oxford: Oxford University Press, 1988.

———. "Knowledge and Power: Health and Physical Education for Women in America." In Rima D. Apple, ed., *Women, Health, and Medicine in America: A Historical Handbook.* New Brunswick: Rutgers University Press, 1992.

———. "Recreating the Body: Women's Physical Education and the Science of Sex Differences in America, 1900–1940." *Bulletin of the History of Medicine* 71 (1997): 273–304.

Vertinsky, Patricia. "Body Shapes: The Role of the Medical Establishment in Informing Female Exercise and Physical Education in Nineteenth-century

North America." In J. A. Mangan and Roberta J. Park, eds., *From 'Fair Sex' to Feminism: Sport and the Socialization of Women in the Industrial and Post-Industrial Eras*, 256–81. London: Frank Cass & Co., 1987.

Vogel, Jeffrey E. "Redefining Reconciliation: Confederate Veterans and the Southern Responses to Federal Civil War Pensions." *Civil War History* 51, no. 1 (2005): 67–93.

Vogel, Morris J. *The Invention of the Modern Hospital: Boston 1870–1930*. Chicago: University of Chicago Press, 1980.

Warner, John Harley. *The Therapeutic Perspective: Medical Practice, Knowledge, and Identity in America, 1820–1885*. 2nd ed. Princeton: Princeton University Press, 1997.

Watson, Richard L. Jr. "A Testing Time for Southern Congressional Leadership: The War Crisis of 1917–1918." *Journal of Southern History* 44, no. 1 (February 1978): 3–40.

Weisskopf, Michael. *Blood Brothers: Among the Soldiers of Ward 57*. New York: Henry Holt and Company, 2006.

Weisz, George. *Divide and Conquer: A Comparative History of Medical Specialization*. New York: Oxford University Press, 2006.

Whalen, Robert Gerald. *Bitter Wounds: German Victims of the Great War, 1914–1939*. Ithaca, NY: Cornell University Press, 1984.

Whitehorne, Joseph W. A. *The Inspectors General of the United States Army, 1903–1939*. Washington, DC: Office of the Inspector General and Center of Military History, United States Army, 1998.

Willrich, Michael. "Home Slackers: Men, the State, and Welfare in Modern America." *Journal of American History* 87 (September 2000): 460–89.

Wilson, Philip D., and David B. Levine. "Hospital for Special Surgery: A Brief Review of its Development and Current Position." *Clinical Orthopaedics and Related Research* 374 (May 2000): 90–106.

Winter, Jay. "Military Fitness and Civilian Health in Britain during the First World War." *Journal of Contemporary History* 15, no. 2 (April 1980): 211–44.

Witt, John Fabian. "The Transformation of Work and Law of Workplace Accidents, 1842–1910." *Yale Law Journal* 107 (March 1998): 1467–1502.

———. *Accidental Republic: Crippled Workingmen, Destitute Widows, and the Remaking of American Law*. Cambridge, MA: Harvard University Press, 2004.

Wood, Ann Douglas. "The War Within the War: Women Nurses in the Union Army." *Civil War History* 18 (1972): 197–212.

Zenderland, Leila. *Measuring Minds: Henry Herbert Goddard and the Origins of American Intelligence Testing*. Cambridge Studies in the History of Psychology. Cambridge: Cambridge University Press, 2001.

Zieger, Robert. *America's Great War: World War I and the American Experience*. New York: Rowman and Littlefield, 2000.

Zola, Irving K. "Medicine as an Institution of Social Control." *Sociological Review* 20 (1972): 487–504.

Index

Page numbers in italics represent photographs and illustrations.

Follett, Albert G., 108–9, 110
Fort McPherson Physiotherapy Department, 72, 74, 161
Foucault, Michel, 40, 205n29
France, Joseph I., 163–64
Freiberg, A. H., 57–58, 59

Gallinger, Jacob H., 163, 164
Gambel, Jeff, 171–72
Gamble, Vanessa, 138
Garrison, Lindley M., 11
gender roles: marriage and family for disabled soldiers, 31, 62, 134; maternalist/paternalist reformers, 30–31, 199n87; and physiotherapists, 63, 68–70, 71, 77–78, 218n74; and the WRIA, 30–31, 62, 199n87. *See also* masculinity/manliness
General Hospital No. 3 (Colonia, New Jersey), 94
General Law of 1862 and Civil War pension system, 13–17, 28–29, 32; and Confederate soldiers, 16–17; means-testing based, 15; Pension Bureau claims, 15, 28, 32, 195n22; Republican presidential administrations' reinterpretation, 17
Georgetown Union Hotel Hospital, 82
Germany: Bismarckian system, 147, 150, 151, 152; industrial safety reforms and workers' health plan, 151; social insurance, 147, 150, 151, 152, 157
Gibney, Virgil, 44, 45–46, 50, 51–52
Gibson, Henry R., 25–26
Giroir, Brett, 174, 175
Gladden, Washington, 19
Glasson, William H., 22–23
Goldthwait, Joel E., 52–60; and Army Division of Orthopedic Surgery, 57–59, 66; and assembly line of standardized care, 52–53, 59; career, 52, 209n86; and disabled soldiers' self-pity and dependency, 223n64; and the "orthopedic war preparedness committee," 54, 56–57; and physiotherapist corps, 63–70, 66, 72, 212n10, 213n11; treatment of working-class adult patients, 52–53; vision of improved postwar society and end to pension system, 57; and wartime military orthopedics, 54–60
Gompers, Samuel, 3, 5, 27–28, 121, 239n20
Gordon, John B., 142, 146

Gorgas, William C.: and Division of Orthopedic Surgery, 5, 36–37, 58, 61; and Division of Physical Reconstruction, 5, 149, 155; executive order to hire "women war workers," 61; expansion of the Army Medical Department, 5–6, 61; and health care for industrial accident victims, 149; and military orthopedics in Britain, 56; military rehabilitation hospitals and the army's commitment to rehabilitative ideal, 79, 80, 149, 155–56. *See also* US Army Office of the Surgeon General (OSG)
Grand Army of the Republic (GAR) and Civil War pension system, 12–13, 16, 23
Grand Review parade (1915), 10–11
Great Britain: industrial safety and National Insurance Act, 151; occupational therapy system and physiotherapists, 63, 213nn11–12; rehabilitation propaganda, 140; social welfare reformers and orthopedic care, 55–56, 63; statistics on soldiers' injuries, 98; US health care system in contrast to, 151, 157; wartime therapeutic massage/massage therapy, 63, 68–69, 213n12
Gritzer, Glenn, 202n10

Hale, William Bayard, 26
Halfaker, Dawn, 179
Hall, John S., 129–30, 131
Hall, G. Stanley, 25
Hanger, J. E. (artificial limb manufacturer), 99, 106–7
Harper's Weekly, 18, 22
Harrison, Benjamin, 18, 20
Harvard Medical School, 52, 57, 66, 101
Hayes, Rutherford B., 17
Hays, Samuel P., 56–57
health insurance: AMA opposition, 150, 165, 243n86; and CVRA legislation mandating vocational rehabilitation, 150; German system, 147, 150, 151, 152, 157; and World War I America, 149–50, 151
Helvétius, Claude, 41
Hendrick, Burton J., 26–27
Heroes All (military instruction film), 76
Herschbach, Lisa, 225n13
Hibbs, Russell, 203n18, 211n115
Hickel, K. Walter, 199n87
Hill, Susan, 71–72
Hirsch, Elihu P., 178

Katz, Michael, 47
Keen, William W., 58
Keene, Jennifer, 102, 137, 226n22
Keller, Helen, 3, 234n33
Keller, William L., 146
Kelly, Patrick J., 15, 224n67
Keogh, Sir Alfred, 56
Kidner, T. B., 158, 160
King, Caroline B., 77–78
Kirkbride, Thomas Story, 94
Knapp, Frederick, 95
Knight, James, 38–40, 47
Koven, Seth, 60

labor and work. *See* industrial economy
    and postwar labor market; vocational
    education
Lane, Jerome E., 129
Lange, Fitz, 36
Lathrop, Julia, 3, 5, 28, 31
Lee, Robert E., 15
Lee Camp Home (Virginia), 16
legislation, Progressive Era. *See* Civilian
    Vocational Rehabilitation Act (CVRA);
    Civil War pension system, legislative
    changes to; Federal Board of Vocational
    Education (FBVE); War Risk Insurance
    Act (WRIA) (1917)
Letterman, Jonathan, 81
Letterman General Hospital (San Francisco),
    80–83, 84–92, 95–97, 156; architecture
    and pavilion design, 82; and Civil War
    pension system debates, 84–85; desegre-
    gated facilities, 87–88, 89, 137; disabled
    noncombatants, 129; the early Letter-
    man Hospital at the Presidio, 81–83; ex-
    panded wartime orthopedic wards, 87;
    hospital newspaper, *The Listening Post*,
    96; orthopedic appliance shop, 89–90;
    physiotherapy building, 90–91; race riot
    and racial tensions, 137; ventilation sys-
    tem, 82; wartime "curative workshop,"
    91–92; wartime rehabilitation goals,
    85–88, 95–96; working-class patients,
    144. *See also* hospitals (military rehabili-
    tation hospitals) of World War I
Lewis, David, 125
Library of the Surgeon General's Office
    (National Library of Medicine), 83
Lindsay, Samuel McCune, 28, 29, 31
Ling, Per Henrik, 68

Little, R. M., 158
Little, William John, 36
Lloyd George, David, 56, 151
Lorenz, Adolf, 36
Los Angeles Times, 10
Lovett, Robert, 45–46, 57, 59, 63, 66
Lusitania sinking (1915), 11

Mack, Julian, 3, 5, 28–32, 122–23, 148
Macon, Nathaniel, 14
Magnuson, P. B., 113, 157
Manoel of Portugal, King, 56
Marble, Sanders, 79–80
Maroney, Mildred, 27
marriage and disabled veterans: and reha-
    bilitation hospitals, 134–35, 235n43;
    WRIA framers' goal of marriage and
    family, 31, 62, 134
masculinity/manliness: and breadwinner
    ideal of manhood, 4; and critiques of
    Civil War pension system, 20–21, 24–26,
    144; fears of emasculating effects of dis-
    ability, 130; female caregivers and moral
    failure of disability, 50; and female war
    workers, 61, 62, 76–77; and military re-
    habilitation hospitals, 95–96; and mili-
    tary rehabilitation propaganda, *127*,
    129–31, 144, 233n15; soldier-patients
    who rejected the image, 131; and "su-
    percrips," 130–31, 180–81; and today's
    amputee-soldiers, 168–69, 180–81; and
    WRIA, 62
Massachusetts General Hospital, 41, 53, 66
Massachusetts Institute of Technology, 173
massage therapy: history and revival of
    massage, 63, 68–69; medicalization and
    professional legitimacy, 74–75; "medical
    rubbers," 70; and physiotherapy aides,
    68–70, 72–75; physiotherapy wards and
    bedside massages, 74; therapeutic inflic-
    tion of pain, 6, 75; training courses,
    69–70
Maverick, Maury, 131, 146–47
Mayer, Leo, 81, 115
Mayo, Charles W., 58
McAdoo, William, 27–28
McMurtrie, Douglas, 47, 62, 65, 75, 140, 141,
    157–60
Medical College of Ohio, 58
medicalization of veteran welfare, 192n36
medical welfarism, 9, 38, 147, 203n19